AF580186

Biological Response Modifiers

Distributed as an educational service
by Cetus Oncology Corporation

Biological Response Modifiers

A Self-Instruction Manual For Health Professionals

Edited by:
Kimberly A. Rumsey, RN, MSN, OCN
Paula Trahan Rieger, RN, MSN, OCN
University of Texas
M.D. Anderson Cancer Center
Houston

Precept Press, Inc., Chicago

Library of Congress Catalog Card Number:
92-81530

International Standard Book Number:
0-944496-30-X

96 95 94 93 92 5 4 3 2 1

Printed in the United States of America

Contents

Illustrations

Tables

Foreword

The last two decades have seen an impressive development in the field of human immunology. The advent of new technologies, such as genetic engineering and generation of monoclonal antibodies, has revolutionized the field to the point that cell subsets are now characterized in detail; and new soluble proteins, which serve as messengers in this system, are being discovered regularly. These rapid developments have not gone unnoticed by the clinicians, and biologic therapy is playing an increasing role in the management of cancer. Cytokines, such as interferons and interleukin-2, have been genetically engineered and are available in abundance for clinical studies. These naturally occurring molecules have already demonstrated activities against various cancers. Hematopoietic growth factors are finding their place as adjuncts to chemotherapy and to bone marrow transplantation, while approaches to expand ex-vivo various cellular components of the immune system are rapidly gaining momentum.

It is as important for nurses as it is for physicians to become experts in these new concepts and to become aware of problems which are specific to the field. Side effects are different and require different management, and parenteral self-administration of these cytokines requires appropriate instruction and direction to the point that the well-trained nurse is emerging as pivotal to the success of a biologic therapy program.

This timely book provides an excellent basic guide for nurses who have an interest in the field of clinical immunology and biologic therapy, as well as practical guidelines for the management of patients treated with biologic response modifiers.

Moshe Talpaz, MD
Deputy Chairman, Department of Clinical Immunology and Biological Therapy
Chief, Section of Biologic Therapy
Associate Professor of Medicine
M. D. Anderson Cancer Center
Houston, Texas

Preface

The primary objective in preparing this book was to attempt to provide comprehensive knowledge about biological response modifiers for nurses who are responsible for the care of patients receiving these agents or who are preparing to assume such responsibilities. The manuscript was prepared especially for use as a teaching tool and informational resource in schools of nursing and in inservice education programs of institutional nursing service departments. In addition, the book should be useful in providing a basic understanding of the components and mechanisms of action of the immune system for other health care professionals engaged in some aspect of clinical care or related clinical research.

An understanding of the immune system is important in the care of patient populations in fields such as oncology, organ transplantation, and immune disorders. Better understanding of the biology of the immune system has resulted in more sophisticated biotherapy, and the discoveries of recombinant DNA technology and hybridoma technology have led to the widespread availability and use of biological agents in clinical trials and, subsequently, in patient care. In preparing this manuscript, the authors hope to contribute to the broadening of familiarity with the workings of the immune system and the advances made in the field of biotherapy.

In three major sections, the book covers the following topics:

- Components and activities of the immune system
- Advances made in the field of oncology related to the use of biotherapy as an anticancer treatment modality
- The theoretical mechanism of action of the various categories of biological response modifiers
- Goals and target population for the three phases of clinical trials
- The nurse's role in informed consent

- How to develop a nursing plan for care of the patient receiving biotherapy based on the specific treatment regimen and expected side effects

In addition, the book includes self-testing materials for readers seeking to measure their grasp of the subject, an annotated bibliography, and an up-to-date listing of informational and teaching resources offered by scientific firms active in the field of biotherapy.

Among the figures and tables augmenting the text are detailed and comprehensive listings of possible side effects of biological response modifiers and nursing strategies for avoiding or minimizing them.

Kimberly A. Rumsey, RN, MSN, OCN
Paula Trahan Rieger, RN, MSN, OCN

Acknowledgments

The authors gratefully acknowledge the contributions to this project by Avi Markowitz, MD, who reviewed the manuscript, and Leon Williams, who assisted with the graphics. They would also like to thank Joyce Alt, Betty Cody, and Cecil Brewer for their support.

About the Editors

Kimberly A. Rumsey is a clinical instructor in immunology at the University of Texas M.D. Anderson Cancer Center, Houston. Ms. Rumsey earned a Bachelor of Science in Nursing degree from the University of Texas Medical Branch, Galveston, and a Master of Science in Nursing degree from the University of Texas Health Science Center, San Antonio. She has worked in the field of biotherapy and addressed the educational needs of nurses caring for biotherapy patients for four years. She is a frequent lecturer on chemotherapy administration and was co-editor of the Oncology Nursing Society's *Biological Response Modifier Guidelines: Recommendations for Nursing Education and Practice.*

Paula Trahan Rieger is a clinical nurse specialist in immunology/chemopharmacology at the University of Texas M.D. Anderson Cancer Center, Houston. Ms. Rieger holds a Bachelor of Science in Biology degree from the University of California at Davis. She received a Bachelor of Science in Nursing degree from the University of California at Los Angeles and a Master of Science in Oncology Nursing from the University of Texas Health Science Center, Houston. She has worked in the field of biotherapy for ten years and has lectured nationally and internationally on this subject. Her past work as an author includes contribution of a book chapter on the subject of biotherapy for nurses.

Contributors

Paula Trahan Rieger, RN, MSN, OCN, Clinical Nurse Specialist in Immunology/Chemopharmacology, University of Texas M.D. Anderson Cancer Center, Houston, Texas

Kimberly A. Rumsey, RN, MSN, OCN, Clinical Instructor for Immunology, University of Texas M.D. Anderson Cancer Center, Houston, Texas

Margaret Harle, RN, OCN, Clinician III, University of Texas M.D. Anderson Cancer Center, Houston, Texas

Continuing Nursing Education Credit

To earn 6.0 hours of continuing nursing education credit, nurse readers should study this book thoroughly and complete the Continuing Education Test (page 125) that is included with the self-testing materials presented in Appendix C. To receive proper credit, respondents should complete the CNE registration form (page 130), which includes a section for recording answers to the test, and return it together with the required fee to the University of Texas M.D. Anderson Cancer Center Nursing Outreach Program, Houston, Texas. A passing score of 70 percent must be attained. **Readers wishing to keep their book intact may photocopy the registration form.**

PART I

The Immune System

Prepared by Paula Trahan Rieger, RN, MSN, OCN; Margaret Harle, RN, OCN; and Kimberly A. Rumsey, RN, MSN, OCN

Historically, anatomy and physiology courses have not covered the immune system in detail. Today, it is important for health care professionals to have an understanding of the immune system in order to care for patient populations in fields such as oncology, transplantation, and immune disorders. Better understanding of the biology of the immune system has resulted in more sophisticated biotherapy, and the discovery of recombinant DNA technology and hybridoma technology has led to the wide availability of biological agents for clinical trials. The purpose of Part I of this volume is to provide the reader with a basic understanding of the immune system, its components, and its mechanisms of action.

Although the immune system is extremely complex, its three major functions are fairly straightforward: defense, homeostasis, and surveillance. The immune system protects the body against bacterial, viral, fungal, and protozoan invasion (defense), removes damaged cells from the body (homeostasis), and recognizes and removes mutated cells from the body (surveillance). This section will review the organs, tissues, and cells involved in the immune system and describe the two subdivisions of the immune system. The first subdivision is innate (nonspecific) immunity, which is the body's first line of defense against foreign invaders. The second is adaptive (specific) immunity, which is activated once an antigen has overcome the innate immune system.

Organs of the Immune System (The Lymphoid System)

The cells responsible for the immune response are organized into tissues and organs collectively known as the lymphoid system. This system can be divided into primary and secondary organs. The primary organs of the immune system include the bone marrow and thymus. The lymph nodes, lymph vessels, spleen, tonsils, and Peyer's patches make up the secondary lymphoid organs. Lymphocytes develop and mature in the primary organs and are stored and activated in the secondary organs (see figure 1).

Primary Lymphoid Organs

The bone marrow is the soft, spongelike tissue found in the cavities of bones. Its major function is hematopoiesis, which is the formation of the cellular blood components such as leukocytes and platelets. The bone marrow is also thought to be the site of B cell differentiation.

The thymus, located behind the sternum, is the site of maturation for T lymphocytes. The thymus is most active during the early years of life; its activity begins to diminish after age 15.

Secondary Lymphoid Organs

Lymph nodes are chains of encapsulated lymphoid tissue found throughout the body. They are located primarily at junctions of lymphatic vessels and are part of a network that drains and filters tissue fluid. They also house both B and T lymphocytes.

The lymphoid vessels form a network that is responsible for draining tissue fluid and returning the transudate to the blood stream. They are also the pathway through which

Figure 1. Primary and Secondary Lymphoid Organs

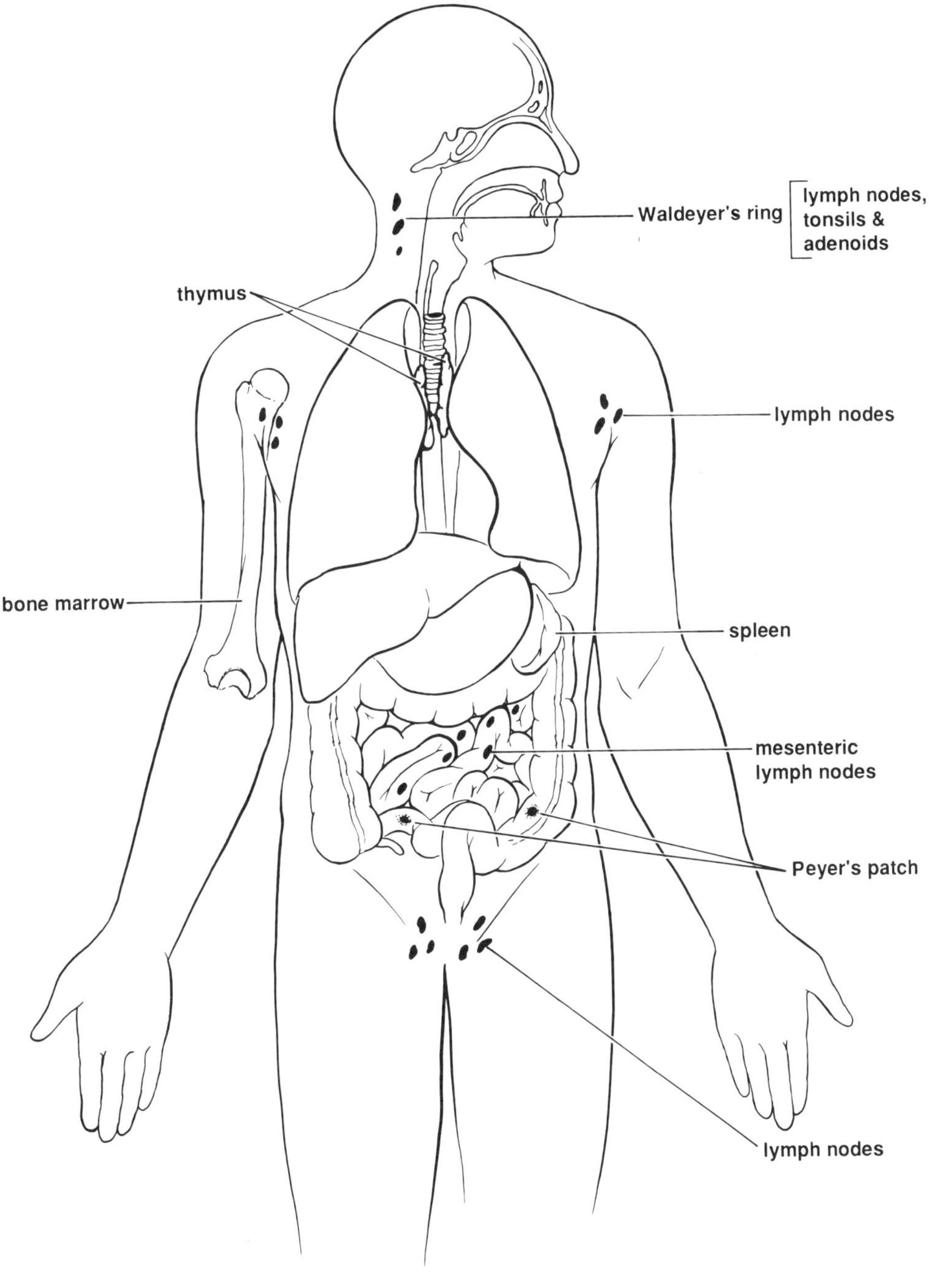

Waldeyer's ring
lymph nodes, tonsils & adenoids
thymus
lymph nodes
bone marrow
spleen
mesenteric lymph nodes
Peyer's patch
lymph nodes

antigens move from the tissues to the lymph nodes for filtration.

The spleen is an encapsulated collection of lymphoid tissue that is located behind the stomach in the upper left quadrant of the abdomen. One of the spleen's functions is to store cells involved in the immune response. Accumulations of nonencapsulated lymphoid tissue, e.g., Peyer's patches and tonsils, are also seen throughout the body. Peyer's patches are groups of lymphoid tissue in the wall of the small intestine. These areas, which are permeable to antigens, house B and T cells. The tonsils, which are located in the pharynx, are arranged in follicles and aid in protecting the body from foreign invasion.

Cells of the Immune System

Like all blood cells, cells of the immune system are formed through a process called hematopoiesis, which occurs in the bone marrow. The pluripotent stem cell, the basic cell of hematopoiesis, has the ability for self-renewal. The stem cell develops into one of two cell lines: the myeloid or lymphoid. The myeloid cell line consists of granulocytes, monocytes, erythrocytes (red blood cells), and megakaryocytes (platelets). The lymphoid line produces T and B lymphocytes. Granulocytes, monocytes, and lymphocytes are primarily responsible for the immune response (see figure 2 and table 1).

Myeloid Cell Line

Granulocytes

Granulocytes are so named because of the characteristic granules that can, with the use of various staining techniques, be microscopically visualized by pathologists. Granulocytes play an important role in inflammation, and

Figure 2. Hematopoiesis

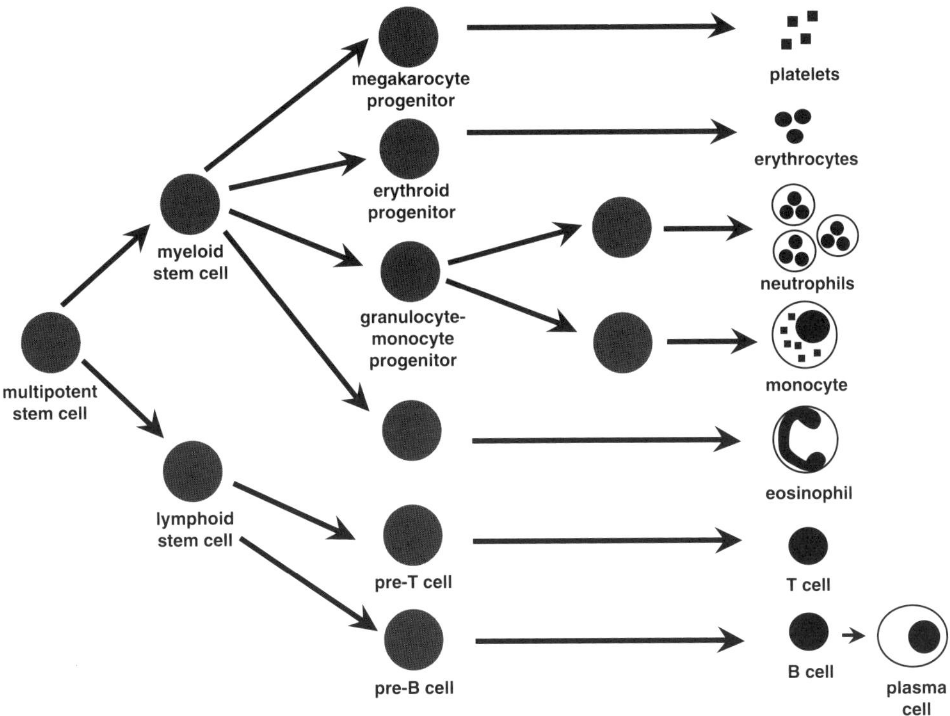

together with antibodies and complement (see "Innate Immunity," below), protect against invasion of microorganisms. They are nonspecific in their action, their primary role being phagocytosis (see "Innate Immunity.") Granulocytes are of three main types:

- *Neutrophils* (polymorphonucleocytes [PMNs]) make up about 50 percent of circulating mature white blood cells. They are stored in the bone marrow and are released as needed to circulate in the blood. After about 12 hours of circulation, neutrophils migrate to the tissues, where they die after about 48 hours. Neutrophils are the main component of the inflammatory response and function primarily as phagocytes to destroy microorganisms that threaten the body.
- *Eosinophils* comprise 1–3 percent of total circulating white blood cells. The exact function of eosinophils is unknown, but they are found in increasing amounts in the circulating blood and in affected tissues during parasitic infections.

Table 1. Normal Blood Values*

Blood counts	Per cu mm	%
Erythrocytes		
Male	5(4.5–6) $\times 10^6$	
Female	4.5(4.3–5.5) $\times 10^6$	
Reticulocytes		
Male		0.8–2.5 (1.6)
Female		0.8–4.1 (1.7)
Leukocytes, total	5,000–10,000	100
Myelocytes	0	0
Band neutrophils	0–700	3
Segmented neutrophils	1,800–6,500	40–60
Lymphocytes	1,000–4,000	20–45
Eosinophils	50–400	1–3
Basophils	0–150	0–1
Monocytes	200–800	4–8
Platelets	150,000–350,000	

*Values may vary according to the laboratory or methods used.

- *Basophils* comprise 1 percent of circulating white blood cells. The large oval granules of the basophil contain pharmacologically active agents that affect the smooth muscle of arterioles and the permeability of capillaries. They are also responsible for early inflammatory changes. When basophils migrate to the tissues during the inflammatory process, they are called *mast cells.* As mast cells, they also generate chemotactic factors. These factors are chemical messengers that attract cells such as eosinophils and neutrophils to the site of inflammation.

Monocytes

Monocytes make up about 2–7 percent of circulating white blood cells and are the largest white blood cell. When they are circulating in peripheral blood, they are called *monocytes.* When monocytes migrate to the tissue, they are then termed *macrophages.* They may live for months at the tissue level. Macrophages are known by different names, depending on the organ in which they ultimately reside:

Organ	Macrophage
Liver	Kupffer's cells
Connective tissue	Histiocytes
Bone	Osteoclasts
Spleen	Dendrite cells
Nervous system	Microglia cells
Lung	Alveolar macrophages

The primary role of the macrophage is phagocytosis. In fact, the literal translation of macrophage is "large eater." The macrophage, one of the primary types of phagocytic cells, is able to recognize and kill invading organisms. Macrophages also play a key role in fighting tumors and in other immune responses.

Lymphoid Line

Lymphocytes

Lymphocytes, which make up about 30 percent of the circulating white blood cells, are located in the lymph nodes, the spleen, and other lymphoid tissue. Lymphocytes are responsible for adaptive immunity. The two major types of lymphocytes are T and B cells:

- *T lymphocytes,* which are responsible for cell-mediated immunity, are produced in the bone marrow and then migrate to the thymus, where they differentiate and mature.
- *B lymphocytes,* which are responsible for humoral immunity, are produced in the bone marrow. The exact site for maturation of the B cell is unknown. However, it has been postulated that maturation occurs in the bone marrow or lymphoid tissue in the gut.

Innate Immunity

Innate immunity, the body's primary line of defense against foreign invaders, is a nonspecific response. The body's first barrier is mechanical and is provided by the skin and mucous membranes, which, when intact, prevent infectious agents from gaining access into tissue, lymphatics, and the blood stream. If the foreign invader penetrates this barrier, other nonspecific mechanisms such as inflammation, complement, phagocytosis, and soluble factors are activated.

Mechanical Barriers (see figure 3)

The epidermis—the external skin layer—is a tough protective barrier covering the entire body. The stratum corneum, the very top layer of cells, normally houses various types of bacterial flora, such as *Staphylococcus epidermis.* However, the skin is an unfavorable environment for other bacteria due to low moisture and antimicrobial substances derived from sweat and sebaceous glands.

An infant has a sterile gastrointestinal (G.I.) tract at birth, but it becomes colonized with bacteria within a few weeks. The body is protected from harmful bacteria by a continuous layer of epithelium, which prevents bacteria from entering the blood stream. Within the G.I. tract, harmful bacteria are destroyed by various means. Saliva and mucus secreted in the small and large intestines both contain lysozyme, an enzyme that kills bacteria. In the stomach, bacteria are destroyed by digestive enzymes, gastric acid, and motility.

In the respiratory tract, small hairs filter large particles and prevent their entry into the respiratory system. The respiratory tract is lined with a ciliated epithelium that moves small particles upward and facilitates their removal. Mucus produced in the respiratory tract helps to trap the particles and prevent dehydration of the epithelium; also, it contains bactericidal lysozyme, which destroys the bacteria. The cough reflex, which is stimulated by inhalation of foreign objects, functions to prevent transport of these objects into the lungs.

Under normal conditions, the bladder, ureter, and kidneys are sterile. The bladder is protected by the antibacte-

Figure 3. Mechanical Barriers

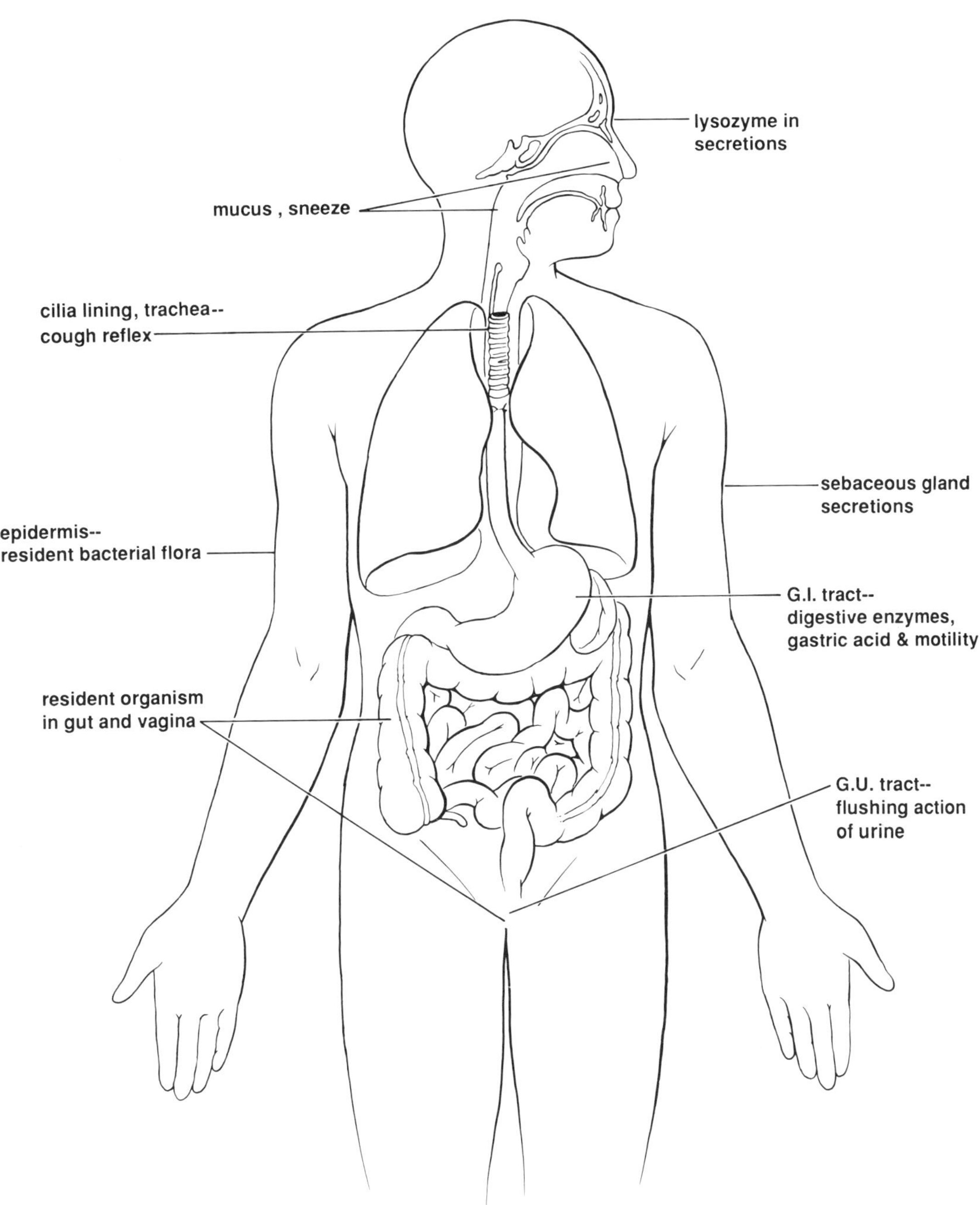

rial properties of the bladder mucosa, and the urethra is protected by the flushing action of urine, which minimizes the ascension of bacteria to the bladder.

Phagocytosis

Phagocytosis is the process by which certain cells of the immune system engulf and destroy invading microorganisms and other antigenic particles. The monocytes, macrophages, and neutrophils—cells that are responsible for phagocytosis—are collectively termed *phagocytes.* When these cells come in contact with an invader, the cell envelops the invader and encloses it within a vacuole. Granules filled with enzymes then fuse with the vacuole, the end result being digestion and destruction of the microorganism (see figure 4). Some phagocytes also play an important role in the immune response by presenting antigen to lymphocytes (see "Humoral Immunity—Antigen-Antibody Reactions," below).

Inflammatory Response

Inflammation is a localized protective mechanism elicited in response to physical injury or invasion by microorganisms. The five cardinal signs of inflammation are redness (rubor), swelling (tumor), heat (calor), pain (dolor), and, occasionally, loss of function. The inflammatory response has both local and systemic components (see figure 5).

Initially, the microorganisms or tissue injury causes chemical mediators (e.g., histamine, bradykinin, or prostaglandin) to be released from the affected tissue, resulting in vasodilation and increased blood flow to the area. These changes are manifested clinically by redness and warmth. As capillary permeability increases, fluids high in protein and fibrinogen shift from the circulatory system to the interstitial space. This results in congestion that may cause

Figure 4. Phagocytosis

phagocytic cell

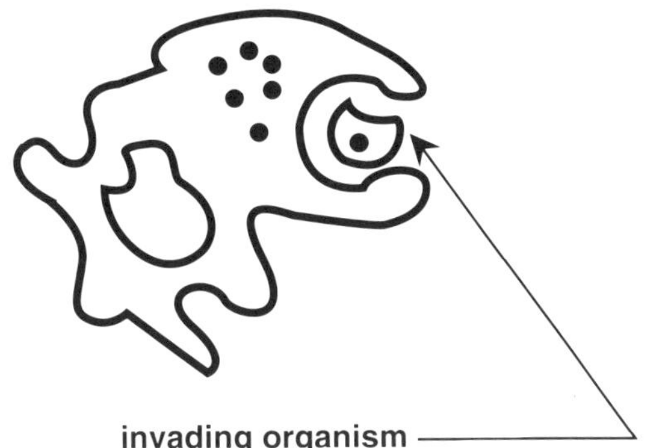

vacuole formation
and fusion with lysome

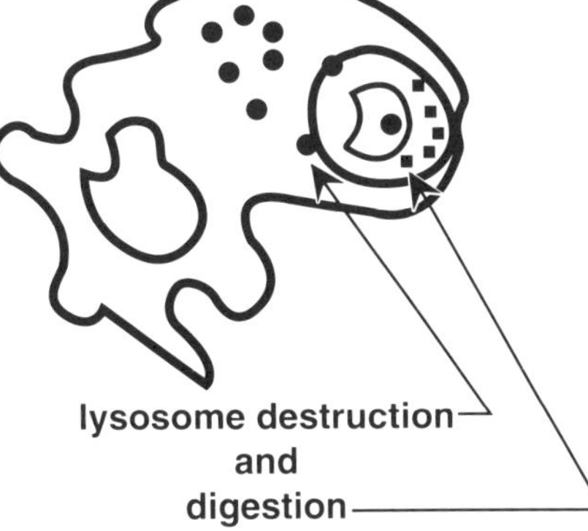

discharge of
undigested products

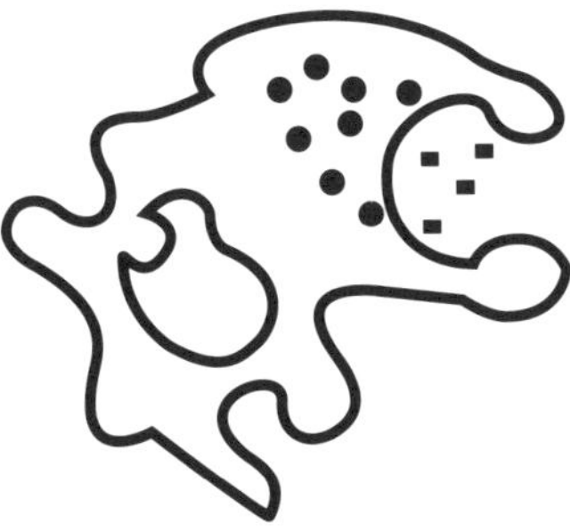

Figure 5. Inflammation Process

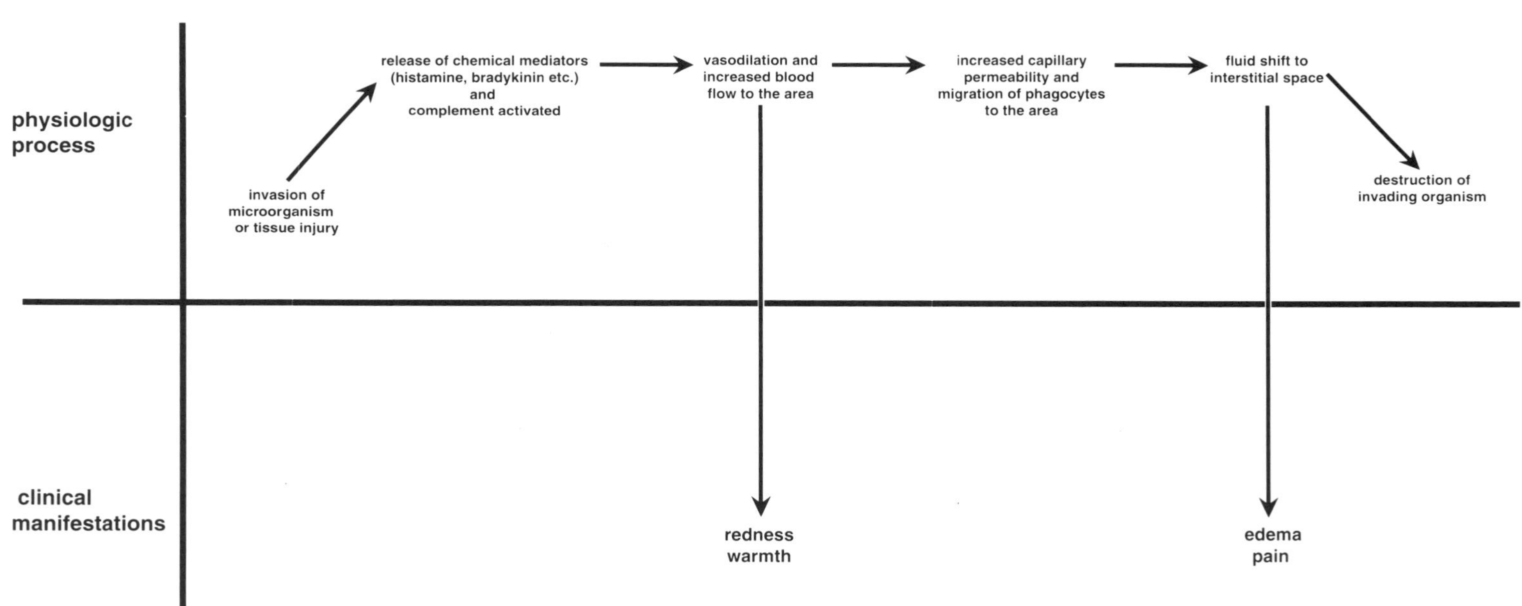

edema and pain. In addition, fibrin clots begin to surround this congested area, further contributing to inflammation. The resulting edema and pain often can lead to reduced function.

Complement and other components of the immune response are also activated, initiating migration of phagocytes into the injured area. The phagocytes then engulf and destroy the invading organism, forming pus. These phagocytic cells also release pyrogens that affect the hypothalmus, resulting in fever. Inflammation causes both a release and an increased production of leukocytes by the bone marrow. The colony-stimulating factors have been identified as one of the major causes of inflammatory leukocytosis. Colony-stimulating factors, acting on the bone marrow, cause cells of the granulocyte and monocyte cell lines to proliferate and mature rapidly.

Natural Killer Cells and Soluble Factors

A variety of soluble factors (e.g., interferon and complement) are involved in mediating the immune response. Communication between different cells of the immune system can be likened to the endocrine system. Soluble proteins produced primarily by immune cells are known as *cytokines.* The primary function of these proteins is immunoregulation. The term *lymphokine* refers to products of lymphocytes, and *monokine* to products of monocytes.

Interferon is one example of a cytokine. The interferons—which actually are a family of glycoproteins—are the first line of defense against many viruses. They are produced by a variety of immune cells and virally infected cells in the early stages of infection. Interferons make uninfected cells resistant to the invading virus and also activate *natural killer* (NK) *cells* (a type of lymphocyte). NK cells represent a subpopulation of lymphocytes. These cells

Figure 6. Natural Killer Cells

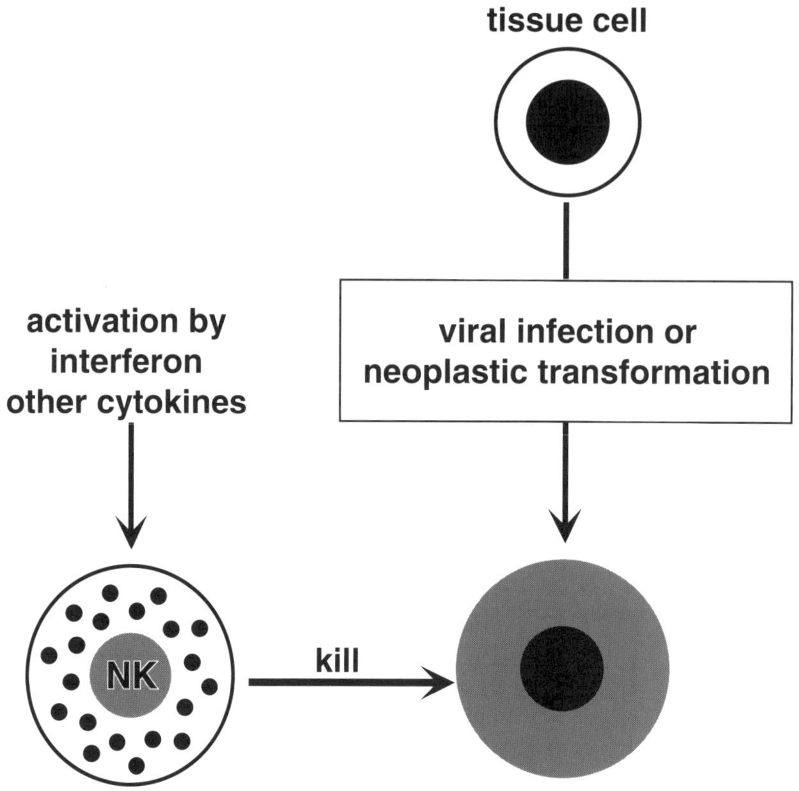

are important in natural immunity in that they recognize virally infected or cancerous cells. Once activated, NK cells identify target structures on the surface of cells, then kill the cell by lysis (see figure 6). These cells are also capable of secreting a variety of cytokines; for this reason, they may have immunoregulatory functions as well as cytotoxic functions.

Complement

Complement consists of about 20 different proteins that interact with each other and with other components of innate and adaptive immunity. These proteins are present in an inactive form in the plasma and other body fluids. Complement is activated spontaneously by the surface of the invading organism. This initiates a "cascade" of sequential reactions in the complement system similar to that in the blood clotting system. Complement helps to mediate the inflammatory process through the following mechanisms (see figure 7):

- *Lysis.* The enzymes digest pieces of the cell membrane, causing cells to rupture.
- *Opsonization.* The complement fragments are deposited on the cell membrane, making it more susceptible to phagocytosis.
- *Chemotaxis.* Certain complement products attract phagocytic cells to the area of inflammation, resulting in an increased number of phagocytes capable of destroying the foreign invader.

Adaptive Immunity

When an individual's innate immunity fails to control a foreign invader, a response known as *adaptive* or *acquired immunity* is called into play. The hallmark of this immune response is its specificity and memory. Specificity refers to the response of the immune cells, which is limited to the antigen for which they are designated. Antigens are substances that can induce a detectable immune response. Adaptive immunity has the ability to recognize foreign antigens and at the same time to tolerate self-antigens. Once an immune response has occurred, if the same organism returns, memory cells that recognize or "remember" the organism will ensure an accelerated and increased response.

The two main types of effector mechanisms in adaptive immunity are *humoral* and *cell-mediated* immunity. The type of reaction initiated depends on the way antigen is presented to lymphocytes. The adaptive immune response can be divided into an afferent limb and an efferent limb. The events of recognition are part of the afferent limb, with the efferent limb being the response phase (see figure 8).

Figure 7. Complement Mechanisms

1. lysis

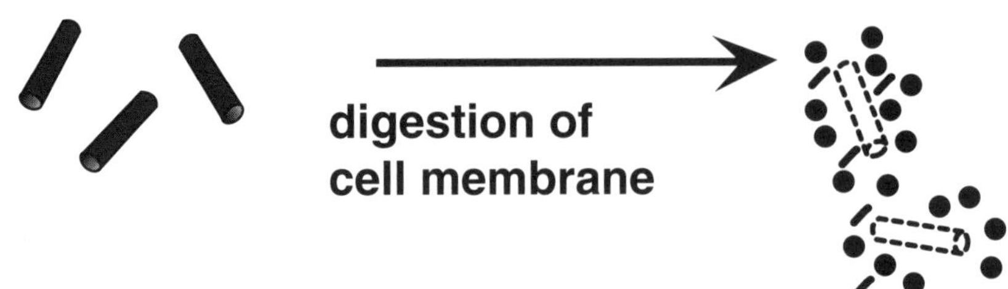

2. opsonization

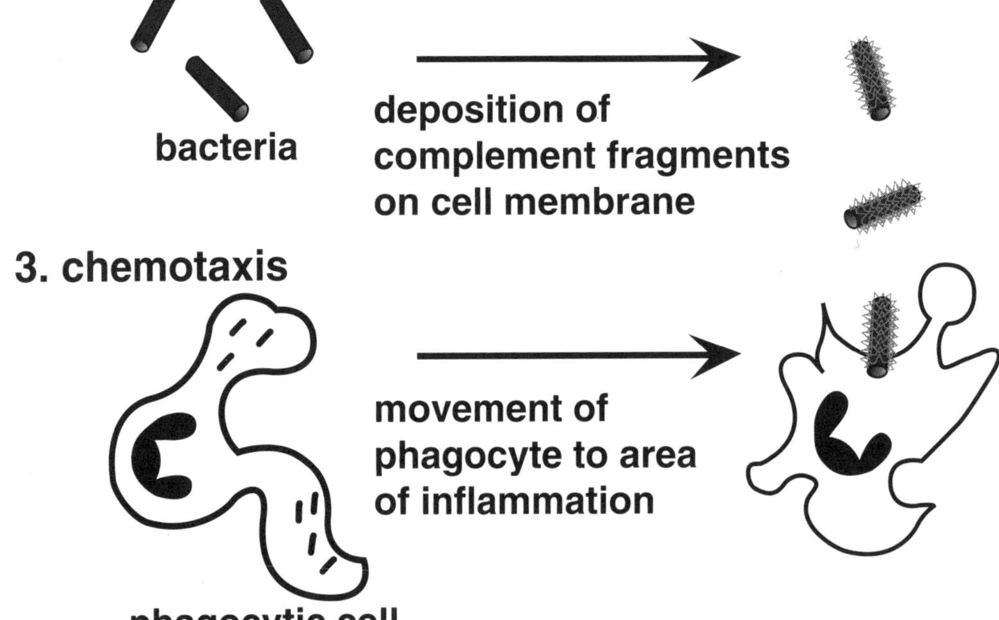

Figure 8. Adaptive Immune Response

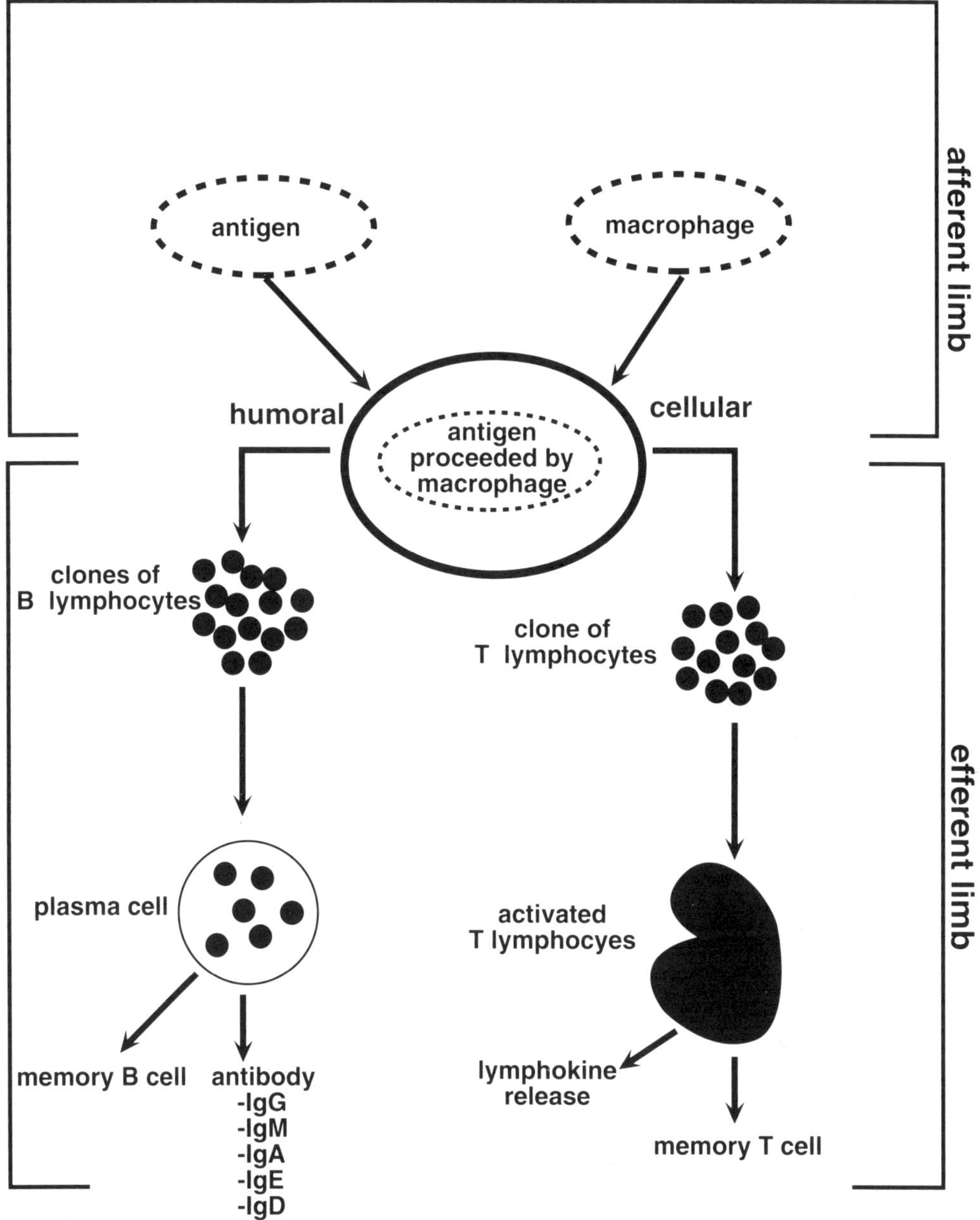

Although the innate and adaptive immune responses perform separate functions, one must remember that interaction does occur between the two.

Humoral Immunity

Humoral immunity protects the body from such pyogenic bacteria as staphylococcus and pneumococcus and from viruses such as influenza, hepatitis, and measles. Humoral immunity is generated by B lymphocytes. The term *humoral immunity* is derived from the fact that B cells generate antibodies that are located in the serum or humoral portion of the blood.

B cells represent about 5–15 percent of the circulating lymphoid pool. They arise from stem cells in the bone marrow and, as they mature, acquire surface markers that identify them as B cells. In birds, B cells mature in a lymphoid organ known as the *bursa of Fabricius,* hence the term *B cells.* In man, however, the bursal equivalent has never been determined. A B cell is defined as lymphocyte capable of producing antibodies.

Antibodies

Antibodies (immunoglobulins) are proteins that are specific for the antigen that induced their production. When stimulated by an antigen, B cells begin to proliferate and further differentiate into plasma cells, the final differentiation product of B lymphocytes. Plasma cells actively secrete antibodies (see figure 8).

Immunoglobulins are glycoproteins composed of heavy and light chains that function as antibodies. Immunoglobulins, located on the surface of B cells, function as specific antigen receptors. When secreted, immunoglobulins serve as effector molecules of humoral immunity.

An immunoglobulin is a unique "Y-shaped" protein (see figure 9). This structure determines the bifunctional capabilities of the immunoglobulin molecule. The top of the Y,

Figure 9. Antibody Structure

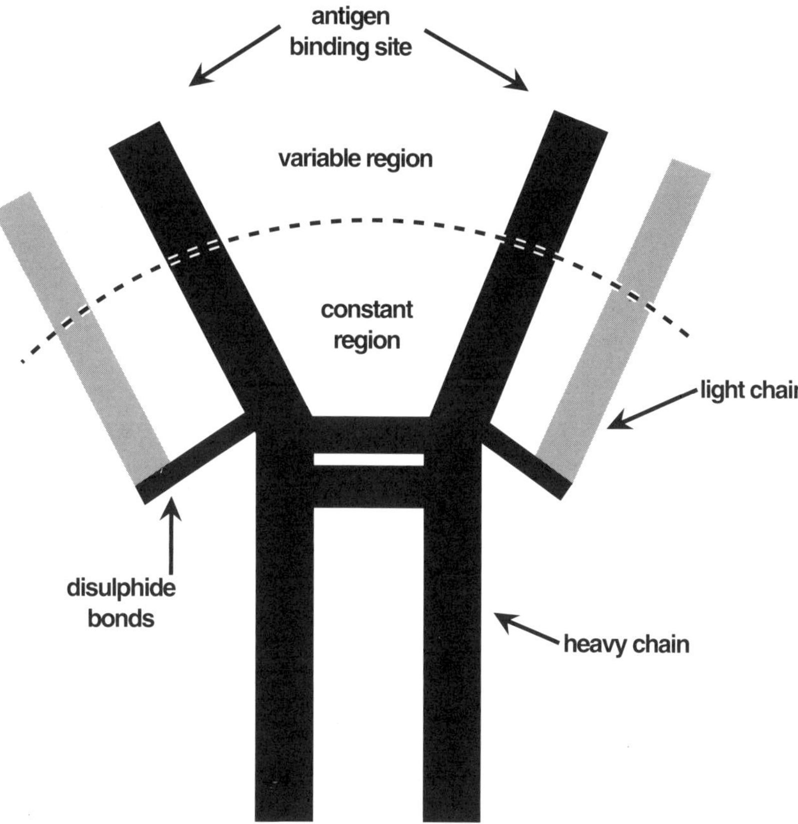

the portion of the molecule that can bind to antigen, is known as the *variable region.* The structure of this region differs among various immunoglobulin molecules. The base of the Y, known as the *constant region,* interacts with receptors on other immune cells.

The immunoglobulin molecule is composed of four polypeptide chains: two identical light chains and two identical heavy chains joined by disulfide bonds. Immunoglobulin molecules are grouped into five classes, which are determined by the type of heavy chain the immunoglobulin has. Each class has different functions and capabilities (see table 2).

Table 2. Immunoglobulins

Class	% of total	Capabilities
IgG	75	Active against most bacteria, some parasites, viruses, and fungi Fixes complement Major antibody of the secondary immune response Crosses placenta
IgA	15	Present in seromucinous secretions (saliva, tears, etc.)
IgM	10	Fixes complement Largest immunoglobulin Efficient agglutination against particulate antigens (RBC, bacteria), transfusion reactions; common antibody to blood group substances Major antibody of the primary immune response
IgE	trace	Hypersensitivity reactions
IgD	<1	Unknown function (?) Membrane receptor on B lymphocytes

Antigen/Antibody Reactions

In general, the activation of B cells and the subsequent antigen/antibody reaction begins when the B cell recognizes an antigen. Immunoglobulins displayed on the surface serve as specific receptors for antigens. To become fully mature (a plasma cell), B cells generally require the assistance of T helper cells. The initial step in this reaction is the processing of the antigen by an antigen-presenting cell (APC). Macrophages often serve this function. Displaying the antigen on its surface, the macrophage presents the antigen to helper T cells. Becoming activated, these helper T cells secrete lymphokines that stimulate B cells to mature into plasma cells, from which antibodies are actively secreted (see figure 10).

Figure 10. Production of Antibodies through Activation of B Cells

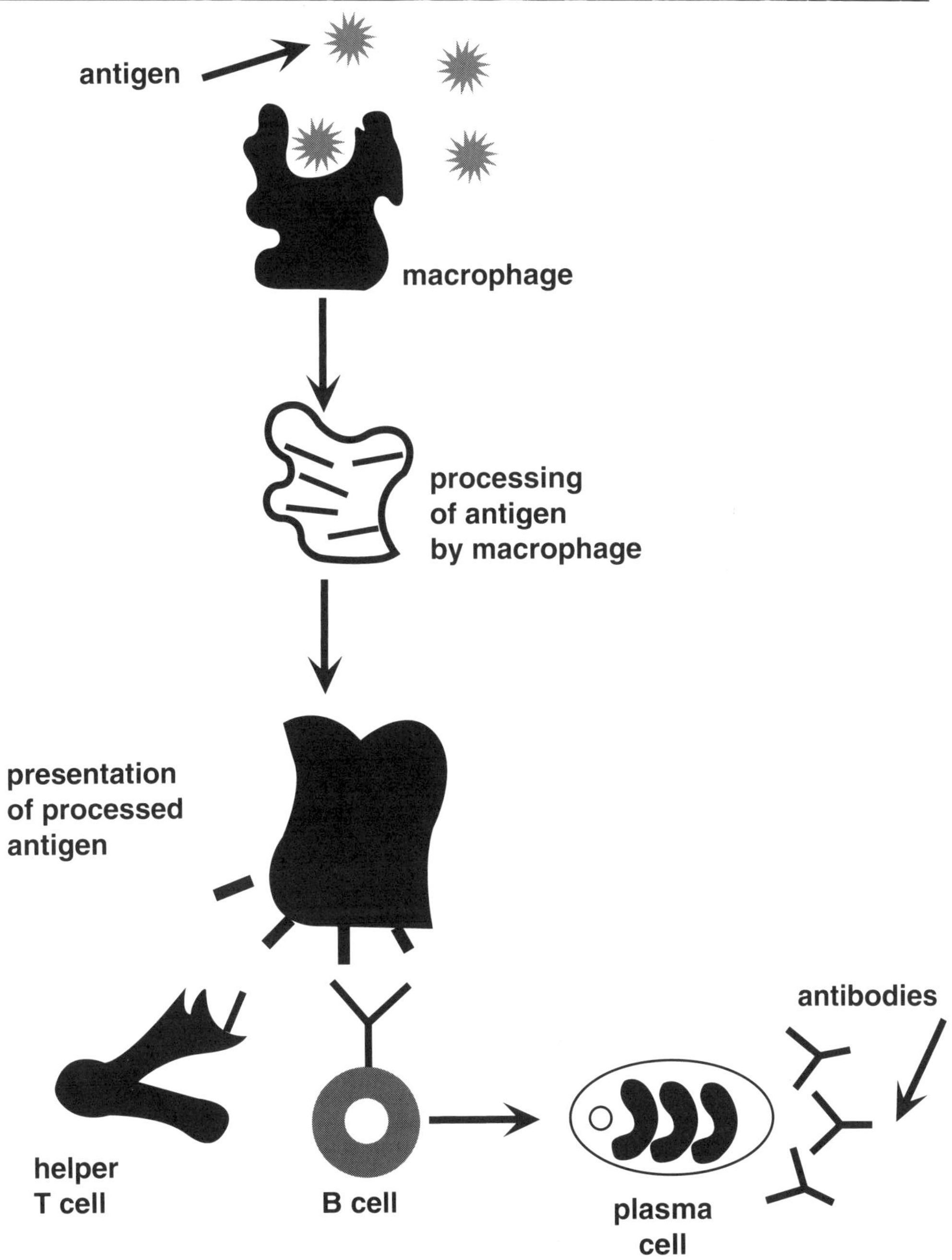

The major function of an antibody is to bind to antigen. The binding of the antigen and antibody can be visualized as the interaction of a lock and key (see figure 11). This binding alone, however, is usually not sufficient to eliminate antigen. The following are usually necessary for elimination of the antigen:

- *Neutralization.* The antibody covers up sites on the antigen, thereby rendering the antigen harmless due to its inability to attach to cells.
- *Opsonization.* The antigen/antibody complex becomes "sticky" and is more easily phagocytized.
- *Agglutination.* The clumping together of antigen-bearing cells in the presence of antibodies.
- *Complement.* Certain classes of antibodies that are capable of fixing complement and eliminating antigen via complement-mediated cell lysis.

Figure 11. Lock and Key Interaction

Primary/Secondary Response

Upon initial encounter with an antigen, the immune system initiates a primary response. At first the response is slow because only a small number of lymphocytes can recognize the antigen, but these cells proliferate to strengthen the immune response. Memory cells are then generated that will recognize the antigen upon subsequent exposure. When reexposed to the same antigen, the immune system initiates a secondary response that is much faster due to the presence of memory cells (see figure 12).

Summary—Humoral Immunity

In summary, B lymphocytes are responsible for humoral immunity, which protects the body from attack by microorganisms. The effector arm of the humoral immune system is antibodies (immunoglobulins) that are specific for the antigen that caused their production. When B cells are activated by an antigen, they generally require the assistance of T helper cells to become fully mature. Known as plasma cells, these mature B cells are capable of secreting large amounts of immunoglobulin. These immunoglobulins bind to and eliminate the foreign invader through various processes.

Cell-Mediated Immunity

Cell-mediated immunity may be defined as immune responses carried out by cells. This type of immunity protects the body against tumors and intracellular organisms such as viruses and parasites and is also responsible for graft rejection. The type of lymphocyte responsible for cell-mediated immunity is the T cell. About 80 percent of circulating lymphocytes are T cells, so named because these cells mature and develop in the thymus gland. Once mature, they migrate to peripheral lymphoid tissue. T lymphocytes may be further divided into subsets according to markers (surface proteins designated CD-4, CD-8, etc.)

Figure 12. Humoral Immune Responses

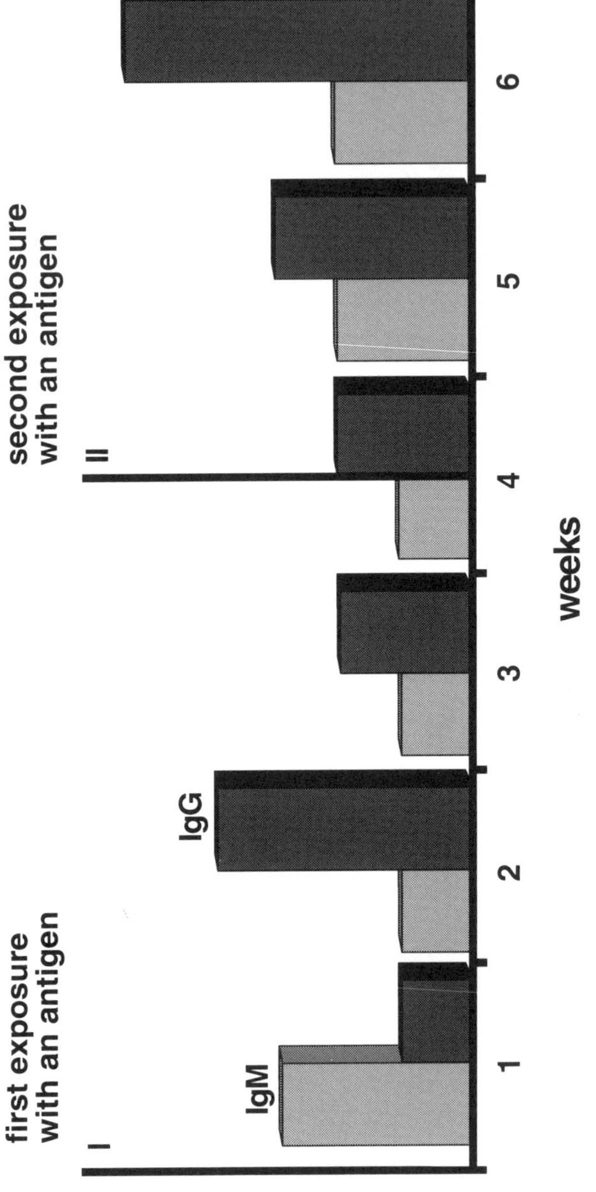

expressed on their cell surface (see figure 13). The main functions of these subsets and their products (lymphokines) are immunoregulation, induction of other immune components, and cytotoxicity. Major subpopulations are:

- *Helper T cells* (CD-4). A functional subclass of T cells that helps initiate the immune response. Activated helper T cells secrete lymphokines, cooperate with B cells in the production of antibody, and help to activate cytotoxic T cells.
- *Suppressor T cells* (CD-8). A subpopulation of T cells that decreases the immune response of other T cells or B cells.
- *Cytotoxic T cells* (CD-8). A subpopulation of T cells that is capable of killing cells by lysis.
- *Memory cells.* Cells with specific antigen receptors. These cells circulate throughout the body in a "resting state" (See figure 8).

Figure 13. Origin of T Cell Subsets

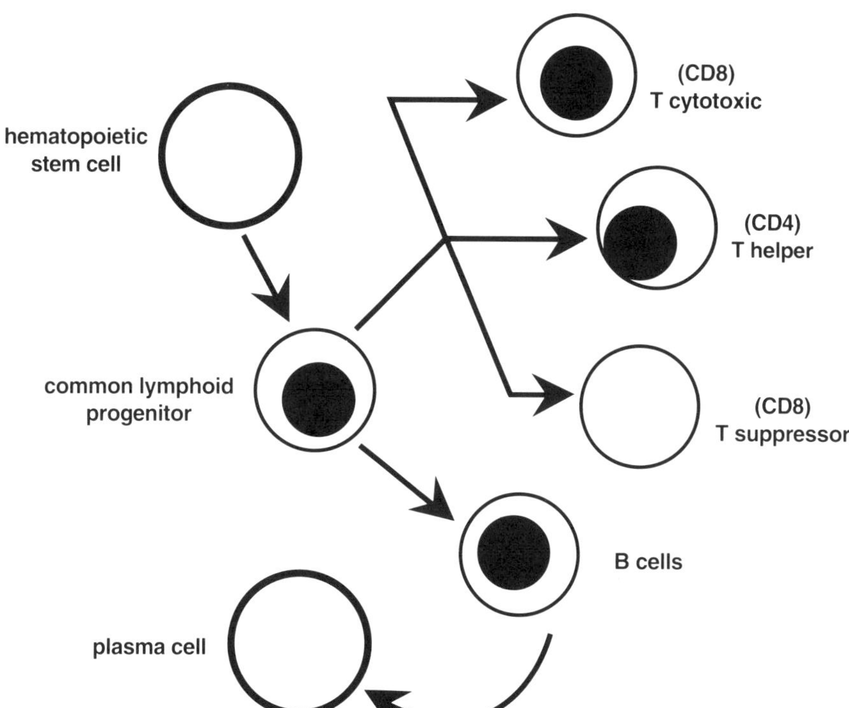

Circulating T cells must be activated by a specific antigen. This process begins when antigen is recognized as foreign by an antigen-presenting cell (e.g., a macrophage) processed and presented to a helper T cell. The helper T cell then binds to the macrophage. This binding, plus stimulation from interleukin-1 (IL-1), a monokine released by the activated macrophage, activates the helper T cell. These activated helper cells then begin to proliferate, secreting additional lymphokines that regulate the immune response. For example, one such lymphokine is interleukin-2 (T cell growth factor or IL-2), which maintains proliferation of activated T cells and stimulates proliferation of cytotoxic T cells. Activated helper T cells also secrete gamma interferon, which enhances cytotoxicity of NK cells. Helper T cells also assist in antibody production by directly interacting with B cells (see "Antigen/Antibody Reactions," above) and by secreting lymphokines such as B cell growth factor (BCGF or IL-4) and B cell differentiation factor (BCDF or IL-6). It is easy to see why helper T cells (CD-4) are crucial in both initiating and magnifying the immune response. (See figure 14 for a summation).

Cytotoxic T cells (CD-8) are a subpopulation of T cells that kill virally infected or cancerous cells by lysis. Unlike NK cells, these cells recognize specific antigen targets. They are also responsible for rejection of foreign tissue and organ grafts.

Another population of CD-8 positive lymphocytes consists of what are known as suppressor T cells. These cells are involved in inhibition, or down-regulation, of the immune response. Although there is much to be learned about how these cells exert their regulatory effects, a variety of suppressor materials have been described. Regulation of humoral immunity can occur by antibody through two major mechanisms. Antibody may combine with antigen and thus competes with antigen receptors of responding B cells. Antibody may also cross-link specific B cell receptors and block the B cells from production of antibody. Regulation of T cell circuits is less well understood. It is hypoth-

Figure 14. Summation of the Immune Response

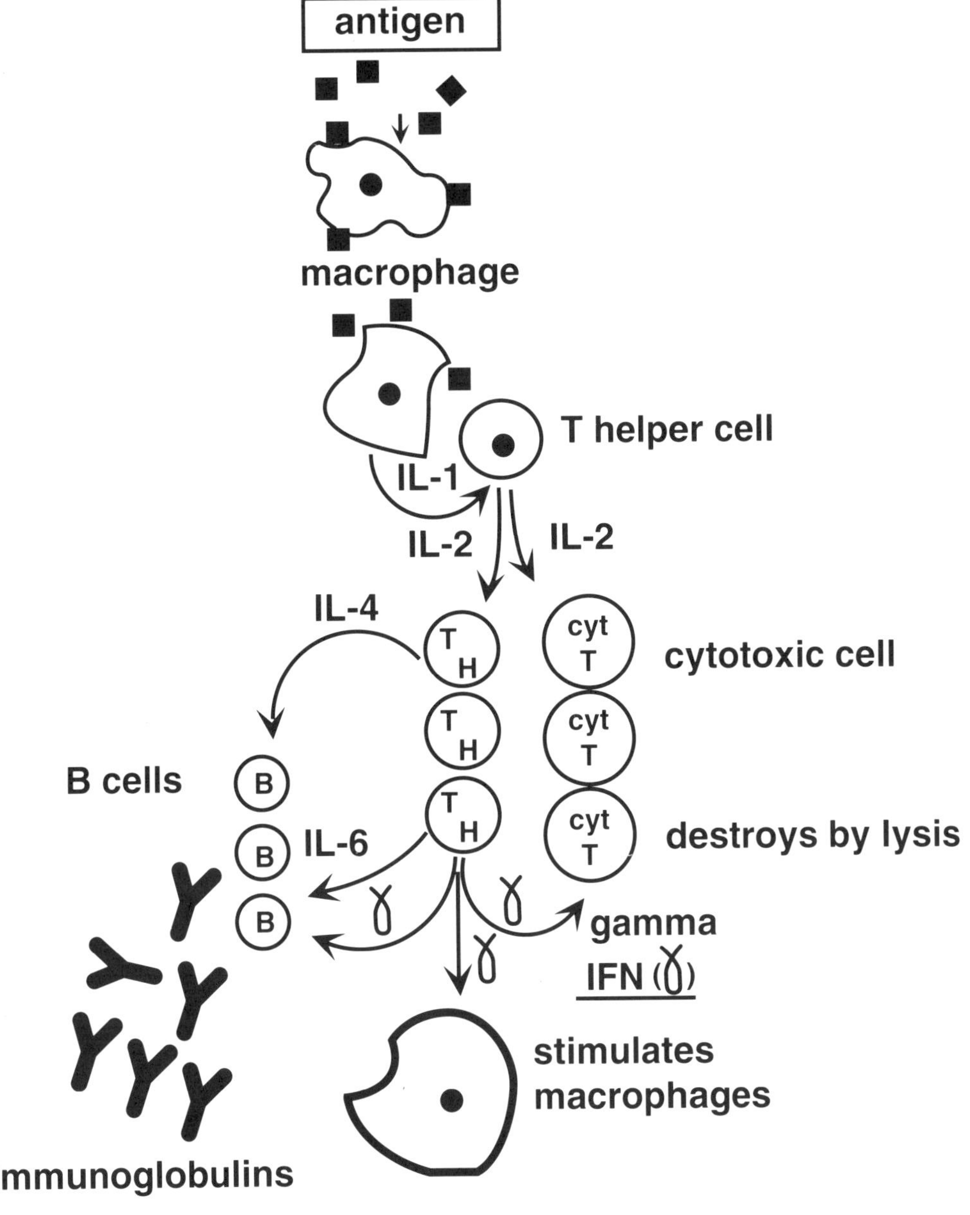

esized that activation of T suppressor cells causes secretion of suppressor factors that ultimately leads to suppression of effector lymphocytes (T and B).

Summary—T Cell Immunity

To review, cell mediated immunity is performed by T lymphocytes. These lymphocytes patrol the body, protecting it from intracellular parasites, destroying virally infected and cancerous cells, and destroying foreign tissues and organ transplants. The four major subpopulations of T cells are: helper, suppressor, cytotoxic, and memory cells. T helper cells recognize antigen processed by antigen-presenting cells and become activated. Through secretion of lymphokines, they direct the activation of other components of the immune system such as cytotoxic T cells or B cells. Regulation of the immune response is partially controlled by T suppressor cells.

Major Histocompatibility Complex

In all mammalian species studied, a single genetic region on a given chromosome is known to code for protein products expressed on the surface of body cells. This genetic region is known as the *major histocompatibility complex* (MHC); in humans it is located on the short arm of chromosome 6. The protein products coded for by this region are sometimes known as histocompatibility antigens.

In humans, this MHC is known as the *human leukocyte antigen* (HLA) *system.* Each gene within this region has multiple alleles. Alleles are one of two or more genes present at a particular site on a chromosome. We inherit one allele from each parent. In the HLA system, two these alleles would then be coexpressed. Thus, each individual has a unique antigen code found on the surface of body cells, due to the expression of HLA alleles. In this way, the immune system is able to distinguish between self and nonself. In an individual's body, proteins (or antigens) expressed on the surface of cells are recognized as self. If these cells are transferred to another person's body, they are recognized by the immune system as nonself.

These cell surface protein products are generally divided into three classes based on tissue distribution. Class I antigens, present on all nucleated cells, serve as the primary targets for cytotoxic T lymphocytes. Class II antigens, which have a more limited distribution, are found on immune cells. Class III genes code for several complement proteins.

The HLA system has multiple loci, including six major antigen groups: HLA-A, B, C, DR, DQ, and DP. Antigens coded for at the A, B, and C loci comprise the Class I antigens, whereas those coded for at the DR, DP, and DQ loci are Class II antigens. In organ and bone marrow transplants, an attempt is made to "match" the alleles expressed in several loci in these major antigenic groups between the donor and the recipient. For example, in bone marrow transplants, the HLA-A, B, and DR antigens are examined, and a donor would be sought who matches the patient in all three loci (six antigens).

Although the HLA system can be very complex, an understanding of the basic principles is crucial for health professionals working in the area of bone marrow transplantation. Knowledge of the three classes of histocompatibility antigens and their associated loci (HLA-A, B, C, DR, DP, DQ) will allow the health professional to assist the patient and family in understanding the confusing process of marrow searches and tissue typing.

Overall Summary

This brief overview of the immune system illustrates the fact that the system is both simple and complex. Although the concepts of what the immune system does may be simple, the manner in which these actions are carried out often are exceedingly complex. The primary functions of the immune system are surveillance, homeostasis, and defense. Two separate types of immunity interact to perform these functions: innate and adaptive immunity. Innate (natural) immunity, the first line of defense, protects the body from foreign invaders nonspecifically. When these barriers fail, adaptive immunity is called into play. The hallmark of adaptive (or acquired) immunity is its specificity and memory. Although scientists now have a tremendous amount of knowledge about the immune system, a great deal remains to be learned. As our knowledge grows, this information is used clinically to understand the pathophysiology of certain immune-related disorders such as AIDS, allergy, autoimmune diseases, and transplantation biology. Many of the cytokines involved in communication between cells and regulating the overall immune response are now cloned for use in clinical trials through the use of recombinant DNA technology. Part II of this volume will explore the use of these agents clinically. (For a list of cytokines, see table 3.)

Table 3. Prominent Biologic Properties of Human Lymphokines/Cytokines

Lymphokine/cytokine	Biologic properties	Produced by
Interleukin-1	Activates resting T cells; is cofactor for hematopoietic growth factors; induces fever, sleep, ACTH release, neutrophilia, and other systemic acute-phase responses; stimulates synthesis of lymphokines, activates endothelial and macrophagic cells; mediates inflammation, catabolic processes, and nonspecific resistance to infection	Activated macrophages (primary source)
Interleukin-2	Is growth factor for activated T cells; induces synthesis of other lymphokines; activates cytotoxic lymphocytes	Activated T cells (helper subset)

Table 3. Prominent Biologic Properties of Human Lymphokines/Cytokines (cont.)

Lymphokine/cytokine	Biologic properties	Produced by
Interleukin-3	Supports growth of pluripotent (multilineage) bone marrow stem cells; is growth factor for mast cells	Activated T cells
Colony-stimulating factors (CSFs)		
Granulocyte-macrophage CSF (GM-CSF)	Promotes neutrophil, eosinophil, and macrophage colonies; activates mature granulocytes	Activated T cells, endothelial cells, fibroblasts
Granulocyte CSF (G-CSF)	Promotes neutrophil colonies	Monocytes, endothelial cells, fibroblasts
Macrophage CSF (M-CSF)	Promotes macrophage colonies	Monocytes, endothelial cells, fibroblasts
Interleukin-4 (B cell stimulating factor-1)	Is growth factor for activated B cells; is growth factor for resting T cells; enhances cytolytic activity of cytotoxic T cells; is mast cell growth factor	Activated T cells
Interleukin-6 (B cell differentiating factor)	Induces differentiation of activated B cells into immunoglobulin-secreting plasma cells	Activated T cells
Gamma interferon	Induces surface antigens on a variety of cells; activates macrophages and endothelial cells; augments or inhibits other lymphokine activities; augments natural killer cell activity, exerts antiviral activity	Activated T cells
Interferon (alpha and beta)	Exerts antiviral activity; augments natural killer cell activity; has fever-inducing and antiproliferative properties	Alpha—primarily leukocytes Beta—primarily fibroblasts

Table 3. Prominent Biologic Properties of Human Lymphokines/Cytokines (cont.)

Lymphokine/cytokine	Biologic properties	Produced by
Tumor necrosis factor (TNF)	Is direct cytotoxin for some tumor cells; induces fever, sleep, and other systemic acute-phase responses; stimulates the synthesis of lymphokines, activates endothelial cells and macrophages; mediates inflammation, catabolic processes, and septic shock	Monocytes/ macrophages (primary source)

Adapted from Dinarello CA, Mier JW. Current concepts: lymphokines. NEJM Oct 8, 1987; 317:15, p. 941.

PART II

Biological Response Modifiers

Prepared by Paula Trahan Rieger, RN, MSN, OCN

For many years, the standard therapies for cancer were surgery, chemotherapy, and radiotherapy. Because of numerous scientific advances over the past decade, biotherapy is becoming accepted as the fourth modality of cancer therapy. This section discusses the rationale for the use of biotherapy, historical perspectives of biotherapy, and the agents currently being investigated. The objective of the material presented below is to acquaint the reader with the classification of biological response modifiers (BRMs) and their mechanisms of action.

Rationale for Use of Biotherapy

The theory of immune surveillance, originally proposed by Burnet in the 1970s, describes a mechanism by which tumors arise in an individual, are recognized as "foreign," and are then destroyed by the individual's immune response. The immune system functions like a watchdog, continuously searching for cancerous cells and destroying them before they grow numerous enough to become tumors. If the immune system becomes depressed or fails to work properly, then cancerous cells may escape detection and consequently develop into a tumor. Although still not totally accepted, this theory has undergone rethinking and modification in recent years. Immunosurveillance may not be confined solely to the development of tumors, but may also play a role in limiting the growth of established tumors or in preventing metastases.

Several clinical examples lend support to this theory. For example, immunodeficient individuals such as radiation victims or organ transplant patients tend to have a higher risk for developing cancer. Radiation victims have a high incidence of leukemia, and organ transplant patients, whose immune response is deliberately suppressed to prevent graft rejection, have a 100-fold higher incidence of lymphoproliferative disorders. Other evidence for this theory rests in histological examination of tumors removed during surgery. Macrophages and lymphocytes found in these tumors may indicate that the immune system had been activated and was attempting to eradicate the tumor. The spontaneous regression of tumors or the regression of metastases after resection of the tumor also suggests that the immune system plays an active role in fighting cancer.

A crucial component of this theory would be the recognition of cancer cells as foreign by the immune system. Research efforts have tried to define the differences between normal cells and cancer cells. A major area of focus has been the identification and characterization of antigens that may be unique to tumor cells and hence recognizable by the immune system. The expression of antigens unique to tumor cells and not seen on normal cells (tumor-specific antigens) has been difficult to prove in human neoplasms. Hence, in humans these antigens are often termed *tumor-associated antigens* (TAAs). Immunologic recognition of neoplasms in humans most likely is due to a quantitative rather than qualitative difference in antigen expression between normal and tumor cells. For example, one such TAA is *carcinoembryonic antigen* (CEA), which is classified as an *oncofetal antigen.* Oncofetal antigens are normally expressed on cells during fetal development and are then reexpressed in large numbers on malignant cells. Another example would be the expression of antigens not found on the normal cell from which a tumor is derived but rather produced by another normal cell type.

Advances in technology have allowed for more sophisticated research into the inner workings of the immune system and its response to malignancy. Many different host responses to tumors have been discovered. It is now known that natural killer cells can recognize malignant cells and destroy them by lysis. Cytotoxic T cells are also capable of recognizing TAAs and killing (also by lysis) the cells that express these antigens. The activation of T helper cells can cause secretion of lymphokines that have an impact on NK cells, macrophages, and lymphokine-activated killer (LAK) cells. In some patients, tumor-specific antibodies have been detected, perhaps indicating that a humoral immune response was initiated.

One problem with the theory of immune surveillance has been the ability to explain why some tumors are able to "escape" detection. There are several hypotheses for why tumors may be able to elude immune defenses. One can imagine a scale on which factors favoring tumor growth and destruction are balanced. Certain factors that may tilt the scale in favor of tumor growth may relate to escape mechanisms. These factors include tumors selected for the ability to escape immune responses, antigenic modulation (shedding, redistribution, or endocytosis), masking of antigens, tolerance, blocking of antigen recognition, and immune dysfunction.

The concept of immune dysfunction is worth exploring. Many factors are thought to affect immune function: age, heredity, nutritional status, obesity, emotional status, and stress. It is known that immune functions are decreased in the very young and the aged and that the ability to generate an immune response is genetically controlled. A variety of nutritional factors are known to influence immunity. All lymphoid organs become smaller during protein-calorie malnutrition, with the thymus being most severely affected. An effect on humoral immunity, however, has not been observed. Vitamin and mineral deficiencies as well as excesses can have an effect on the functioning of the immune system. Limited experiments conducted thus far have shown that any vitamin involved in the synthesis of DNA and protein has an effect on the immune system. In general, vitamin deficiencies appear to have the greatest impact on humoral immunity. Examples include deficiencies of vitamin B_6, vitamin A, biotin, and folic acid. Vitamin C is needed for the maintenance of epithelial cells and the functioning of polymorphonucleocytes (PMNs). Iron deficiency is associated with high infection rates due to its role in the maintenance of lymphoid tissue and the bactericidal activity of PMNs. Massive doses of certain vitamins—A and E, for example—can be immunosuppressive.

One of the most interesting areas being explored today is that of the mind/body connection. Research is now being focused on the effects on the immune system of behavior, emotional status, social interactions, and stress. This new field, known as psychoneuroimmunology (PNI), may be defined as the study of the measurable interaction between psychological and immunological processes. *Psycho* refers to cognitive and emotional processes and mood states; *neuro* to the neurological and neuroendocrine systems and secretions, and *immunology* to the adaptive and innate immune responses. The idea of a mind/body connection is not new; it dates as far back as the sixth century. However, modern medicine and physiology generally has scoffed at such ideas. Emergence of research over the past 10–20 years proving a connection between the immune system and the central nervous and endocrine systems has led to a reexamination of these ideas. PNI serves as a model for examining an organism's psychobiological adaptation to its environment and the influence of psychosocial factors on health.

Researchers have discovered that neurological, endocrine, and immunological cells share biochemical messengers and membrane-associated receptors that allow them to "talk" to each other. These interactions, which occur at many levels, are not yet completely understood. The picture that begins to emerge is that of closely interlocked systems that facilitate a two-way flow of information through these chemical messengers.

Stress can serve as a paradigm for elucidating these concepts. Numerous studies have documented the negative effects of stress (e.g., bereavement, academic examinations) on the immune system. PNI provides a model of how this occurs physiologically. The biochemical cascade occurs along the hypothalamic/pituitary/adrenal axis. Under stressful conditions, the hypothalamus produces a chemical known as *corticotropin-releasing factor* (CRF), which in turn stimulates the pituitary to secrete adrenocorticotropic hormone (ACTH). Stimulation of the adrenal glands by ACTH causes the release of steroid hormones (glucocorticoids) into the bloodstream. The deleterious effects of glucocorticoids are well known: decrease in circulating T cells, selective depletion in lymphocyte subpopulations, inhibition of macrophage differentiation and monokine production, and altered NK cell activity.

Emotions are also thought to affect immune function. The limbic system, made up of a group of subcortical structures in the brain (hypothalamus, hippocampus, and amygdala) is concerned with emotion and motivation. Neuropeptide receptors are found in abundance on cells of these structures as well as on cells of the immune system, suggesting a direct link between the limbic system and the immune system.

The relationship between cancer and psychological variables such as life stress, health habits, and personality traits has long been of interest. Although studies have been conducted evaluating the relationship between psychological factors and immune functioning, they are generally not well controlled and are retrospective in nature. PNI certainly provides a framework for future research evaluating the relationship between cancer onset and progression and the functioning of the immune system. The more intriguing question may then become whether psychotherapeutic or behavioral interventions can directly enhance immune functions and thereby prevent or alter the course of disease. Research in this area remains in its infancy.

The rationale for using biotherapy as a therapeutic approach, then, rests on the tenets of the theory of immune surveillance. Although many questions remain to be answered, it is certain that there is an interaction between the immune system and cancer. Efforts to attack cancer through exploiting these interactions center on stimulating or replenishing the patient's biological responses with the group of substances known as biological response modifiers.

Historical Perspectives

Some of the earliest immunotherapy trials were started early in this century by William B. Coley, M.D. He noticed that patients who suffered severe infections after surgery for malignancy seemed to have a longer tumor-free interval. To induce an infectious response in his patients, he first tried administering live bacteria; later he used filtered toxins. These toxins, which became known as *Coley's toxins,* were used clinically until the 1970s. It is hypothesized that these toxins may have acted to boost the immune system to recognize and destroy cancer cells.

The concept of immunotherapy gained popularity in the late 1960s and early 1970s. Most of the trials involved administration of various bacterial extracts to nonspecifically boost the immune system. Examples of such agents were bacille Calmette-Guerin (BCG), *Corynebacterium parvum (C. parvum),* and other mixed bacterial vaccines. Some controlled studies yielded positive results, but most yielded marginal or negative results.

The failure of immunotherapy to establish itself as a major treatment modality occurred for a variety of reasons:

1. Immunotherapy was more effective for small tumor burdens.
2. Definition and purity of immunotherapeutic agents was lacking.
3. Experimental procedures varied widely.
4. Animal and human models were dissimilar.

Major technological advances of the late 1970s and 1980s have placed the field of biotherapy firmly in the forefront. Because of these advances, biological control mechanisms are now envisioned on a much broader base than the immune system—hence the term *biotherapy* has emerged. In its broadest sense, biotherapy refers to agents derived from biological sources and/or use of agents that affect biological responses. Primarily, these are agents of the mammalian genome. The term *biologicals* describes agents extracted from or produced from biological material. BRMs are agents or approaches that use mechanisms of action involving the patient's own biological responses.

Several important technological advances have brought us to this point. First, an increased understanding of the biology of the immune system has solidified the rationale for this form of therapy and has led to the discovery of new agents with potential clinical application. Second, recombinant DNA technology has allowed for the production of large quantities of highly purified proteins by placing human genes inside bacteria or yeast cells. Third, the discovery of hybridoma technology made possible the production of large quantities of very pure antibodies known as *monoclonal antibodies.* Finally, expanded technology has allowed for the growth of certain immune effector cells in culture, thus allowing them to be administered to patients.

Many clinical trials are now in progress to evaluate the effectiveness of BRMs. Unlike the early immunotherapy trials, in which agents were effective only in patients with "minimal disease," current trials indicate that BRMs are effective in patients with clinically apparent disease. Future trials will attempt to define which agents are most effective and which tumors are most responsive to those agents. Clinical trials should look at the pharmacokinetics and biological responses affected by the agents used. As the knowledge base continues to develop, scientists will strive to develop better preclinical models that can predict the effectiveness of agents.

Classification of Agents

Because of the wide variety of agents currently under investigation, it is at times difficult to understand how BRMs are classified. In general, BRMs can be classified by mechanism of action into three major divisions:

- Agents that augment, modulate or restore the host's immune responses
- Agents that have direct antitumor activity
- Agents that have other biological effects (differentiation agents, agents that affect the tumor's ability to metastasize, or agents affecting cell transformation).

However, it is important to keep in mind that the mechanism of action of many agents is still not completely understood and that many agents have more than one antitumor mechanism. Interferon, for example, acts as an immunomodulator and also has direct antitumor activity in that it slows tumor cell growth. Examples of current biotherapeutic agents and their proposed mechanisms are shown in table 4.

Table 4. Biologicals and Biological Response Modifiers

Agents	Mode of action*
Interferons	
Alpha, beta	I, D
Gamma	I, D
Cytokines/lymphokines	
IL-1	I, B
IL-2	I, B
IL-4	I, B
Tumor necrosis factor	I, D
Colony-stimulating factors	
GM-CSF	B
G-CSF	B
M-CSF	B
Multi-CSF (IL-3)	B
Effector cells	
LAK, TIL	D
Monoclonal antibodies	I, D Diagnostic
Tumor-associated antigens (vaccines)	I

*I = immunomodulatory
B = other biological effects
D = direct effects

Modes of action listed are major presumed modes of action. Table adapted from Abernathy E. Biotherapy: an introductory overview. Oncol Nurs Forum (Suppl) 1987; 14:6.

Agents

This section defines and discusses BRMs currently under investigation and those that have been approved by the Food and Drug Administration (FDA). Also, the mechanism of action and clinical effectiveness of each will be discussed. (A discussion of related side effects and nursing management will follow in Part III.)

Interferons

Interferon (IFN) was identified in 1957 by Isaacs and Lindemann as a soluble substance capable of protecting cells from attack by virus. It is now known that IFNs are a family of glycoproteins consisting of three major antigenic types. The IFN originally derived from leukocytes is known as *alpha* (α). IFN derived from fibroblasts is known as *beta* (β). The third IFN, *gamma* (γ), is derived from T lymphocytes. All IFNs are now produced using recombinant DNA technology.

The effects of IFN begin with its binding to a cell surface membrane receptor. IFN-α and IFN-β share the same receptor, whereas IFN-γ has its own receptor. Upon binding to the receptor, IFN is internalized and then degraded (see figure 15).

The biologic properties of IFNs are not entirely understood, but they are known to have antiviral, antiproliferative, and potent immunomodulatory effects. Each of these actions alone—or several in concert—may be responsible for the anticancer effects of IFN.

Of all BRMs recombinant alpha IFN (rIFN-α) was the first one approved for clinical use by the FDA. In 1986, IFN-α was approved for the treatment of hairy-cell leukemia (HCL). In 1988, high-dose IFN-α was approved for the treatment of Kaposi's sarcoma. Alpha IFN has also been approved for treatment of condyloma acuminatum and

Figure 15. Receptors for Interferons

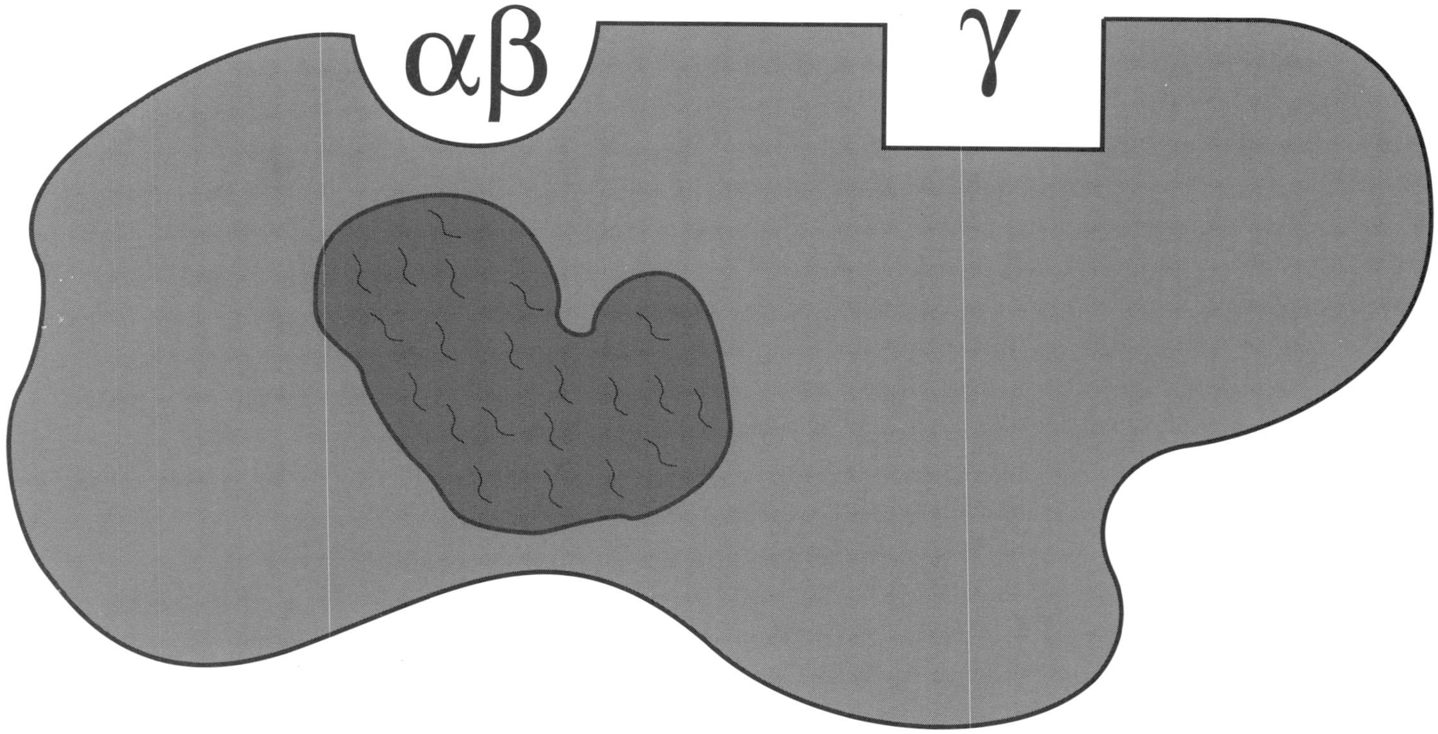

αβ

γ

hepatitis C. IFN-γ recently was approved for treatment of chronic granulomatous disease.

Clinical trials with IFN started in the late 1970s and continue today. IFN, both alone and in combination with other BRMs or chemotherapy, is being tested on a wide variety of malignancies. To date, most clinical successes with IFN have been seen in the hematologic malignancies. Examples are HCL, chronic myelogenous leukemia (CML), chronic lymphocytic leukemia (CLL), multiple myeloma, and low-grade lymphomas. Response rates in the range of 10–20 percent have also been seen in solid tumors such as renal cell carcinoma and melanoma. Recently, studies have shown improved response rates in metastatic colorectal carcinoma for the combination of 5-fluorouracil (5-FU) and rIFN-α as compared to single-agent 5-FU. Future trials will focus on clarifying the role of this combination by comparing it to single-agent 5-FU in randomized trials. In addition, future trials should investigate alternate schedules that would improve efficacy and reduce toxicity.

Although successes with IFN therapy have been seen, a great deal of work remains to be done. The best dose, schedule, route of administration, and type of IFN to use remain to be determined.

Interleukin-2

Interleukin-2 (IL-2) is a lymphokine that was first termed *T cell growth factor.* Produced primarily by activated T helper cells (CD-4), it has proved to be a potent immunomodulator with a variety of actions:

- Stimulates growth and maturation of T-cell subsets
- Stimulates cytotoxic T cells
- Stimulates production of other lymphokines/cytokines.

Exactly how these actions exert an anticancer effect is unknown.

Recombinant IL-2 is now approved for the treatment of metastatic renal cell carcinoma, with investigation ongoing in various other disorders such as melanoma, hematologic malignancies, and infectious diseases. Over the past six years, IL-2 has undergone extensive investigation alone and in combination with lymphoid effector cells such as lymphokine-activated killer (LAK) cells, with tumor-infiltrating lymphocytes (TILs), or with other BRMs or chemotherapy.

The initial clinical work with IL-2 and IL-2/LAK was carried out by Rosenberg at the National Cancer Institute. Some encouraging responses have been seen in patients with renal cell carcinomas and melanoma. However, the full potential of this potent lymphokine remains to be determined. Currently, randomized clinical trials are evaluating the efficacy of IL-2 as single-agent therapy versus IL-2 plus LAK cells. A second active area of investigation is the combination of IFN-α and IL-2 for the treatment of renal cancers and melanoma. One challenge with IL-2 will be to find optimal dosing schedules that enhance efficacy and limit toxicities. Noticeable progress has already been made in this area.

Tumor Necrosis Factor

Tumor necrosis factor (TNF), a monokine produced by activated macrophages, is capable of directly killing tumor cells by inducing hemorrhagic necrosis. TNF therefore appears capable of destroying malignant cells but not normal cells. Laboratory studies have shown TNF to be the same molecule as cachectin, a macrophage-derived polypeptide that appears to be responsible for the metabolic wasting associated with chronic infections or cancer and the early changes associated with septic shock. Now that TNF is available in recombinant form, a variety of phase 1 trials have been conducted across the nation.

The exact mechanism by which TNF exerts its anticancer effect remains to be determined. It is known to bind to receptors on cell membranes and is then thought to be internalized. It has been shown to be both cytotoxic and cytostatic to tumor cell lines in vitro. It appears to cause cell death through necrosis and does not harm nonmalignant cells. Emerging data show that TNF is capable of stimulating the release of IL-1, serves as a growth factor for fibroblasts, and induces the expression of cell surface antigens. These and other yet-to-be-discovered biological effects may contribute to its anticancer effect.

Trials with TNF in animal models have shown it to be a very potent anticancer agent. As yet, trials in patients have yielded disappointing results. Current trials are focusing on the use of TNF in combination with other BRMs and chemotherapy.

Monoclonal Antibodies

In 1975, two immunologists, Köhler and Milstein, developed a technique that allowed production of large quantities of pure immunoglobulin (monoclonal antibodies). The technique is known as hybridoma technology, for which Kohler and Milstein were awarded a Nobel prize. A hybridoma is a somatic cell hybrid formed by fusion of normal lymphocytes and tumor cells; the resulting hybridoma cells will produce the same secretion as the normal parent cells and proliferate indefinitely in culture like the parent tumor cells. B cell hybridomas are formed by the fusion of antibody-secreting B lymphocytes and myeloma cells and are used in the production of monoclonal antibodies. T cell hybridomas, formed by the fusion of T lymphocytes and cancer cells, are useful in the production of T lymphocyte-derived lymphokines in vitro.

Monoclonal antibodies (MoAbs) are produced using this technique and are currently an active area of investigation. MoAbs, which are very specific antibodies produced from a single clone, are directed against a single antigenic determinant on the surface of a cell. MoAbs have been produced for a wide variety of cancers. Antigenically complex cancers such as melanoma may be targeted by a large number of different MoAbs.

MoAbs have a wide variety of potential applicability in the diagnosis and treatment of cancer. They can be used in medical laboratories to improve understanding of cell differentiation and cancer biology. They are of great value to the pathologist in the histopathological classification of cancers. When low-dose radioisotopes are attached to MoAbs, they can be used for imaging tumor masses. They might be used to target therapy such as chemotherapy, biotherapy, or radiotherapy to the tumor. Some MoAbs are capable of inducing cell membrane lysis directly through interaction with other immune cells or complement. MoAbs are also being evaluated for purging autologous bone marrow of cancer cells before transplantation or for the selective removal of T cells responsible for graft-versus-host disease from marrow prior to allogeneic transplant.

The next few years look exciting for the study of MoAbs. The majority of trials at this time use murine MoAbs. However, as research moves forward, we will see more clinical trials using human MoAbs or chimeric MoAb (part murine/part human).

Colony-Stimulating Factors

Hematopoiesis is the process by which cells of the hematopoietic lineage are produced (see "Cells of the Immune System" in Part I). The process starts with a pluripotent stem cell that differentiates into progenitor cells that are capable of maturing into different hematopoietic lines. A class of proteins known as *colony-stimulating factors* (CSFs) has been shown to support this process in vitro. Strong evidence has accumulated that the physiological role of the CSFs apparently is to regulate the circulating concentrations of granulocytes and mononuclear phagocytes. Hence, their most direct role in host defense relates to controlling the number of effector cells. CSFs are also responsible for modulating functions of mature effector cells.

CSFs are named primarily for the lineages on which they have an impact (see below).

The CSFs, currently an area of active research, are produced by a variety of immune cells. Several CSFs have been cloned, and clinical investigations continue. Areas in which CSFs may have potential clinical applications include:

- Decreasing myelosuppression secondary to chemotherapy/radiotherapy
- Accelerating recovery from autologous bone marrow transplantation
- Benefiting patients with pancytopenia due to aplastic anemia or myelodysplastic syndrome
- Treating infections and parasitic diseases.

Granulocyte-Macrophage Colony-Stimulating Factor

Granulocyte-macrophage CSF (GM-CSF) is available in recombinant form and was approved by the FDA early in 1991 for the treatment of neutropenia after bone marrow transplantation in patients with non-Hodgkin's lymphoma (NHL), acute lymphoblastic leukemia (ALL), and Hodgkin's disease. Active investigation continues in a variety of clinical trials, and further approved applications are anticipated. GM-CSF primarily stimulates the production of neutrophil and monocyte colonies. It has also been shown to augment in vitro growth of erythroid progenitors and eosinophils. Its effects on platelets and lymphocytes are unclear.

Studies have shown that GM-CSF has a number of effects on the functions of mature neutrophils:

- Inhibits migration of neutrophils in vitro
- Primes neutrophils for enhanced oxidative metabolism
- Augments capacity of neutrophils to phagocytize
- Induces macrophage tumoricidal activity.

Current studies will continue to evaluate the use of GM-CSF after chemotherapy or in patients with primary or secondary bone marrow failure. Future studies will further refine and clarify the role of GM-CSF and define its use in combination with other CSFs or BRMs.

Granulocyte Colony-Stimulating Factor

Granulocyte CSF (G-CSF) is a neutrophil-specific hemopoietin. It stimulates the growth of neutrophil colonies and affects mature neutrophil functions. For example, G-CSF is known to enhance phagocytic activity of neutrophils and antibody-dependent killing. Cells producing G-CSF are monocytes, fibroblasts, and endothelial cells. G-CSF is produced in recombinant form, and in early 1991 received FDA approval for treatment of neutropenia in patients developing nonmyeloid malignancies after antineoplastic therapy. G-CSF is generally started a minimum of 24 hours after chemotherapy and is continued until the *post nadir* absolute neutrophil count (ANC) reaches about 10,000/mm^3. G-CSF is also used to accelerate neutrophil recovery after bone marrow transplant.

Experimental studies have shown that G-CSF can cause the egress of hematopoietic progenitor cells into the peripheral blood. Currently, clinical trials are evaluating the use of stem cells harvested from the peripheral blood after G-CSF therapy to repopulate the bone marrow following aggressive chemotherapy.

Macrophage Colony-Stimulating Factor

Macrophage CSF (M-CSF) is a macrophage-specific growth factor. M-CSF is produced by monocytes, fibroblasts, and endothelial cells. M-CSF stimulates monocyte progenitor cells and has prominent effects on the functioning of mature monocytes/macrophages. Recombinant M-CSF is available, and human clinical trials are currently in progress.

Interleukin-3 (IL-3)

Interleukin-3, also known as multi-CSF, is an early growth factor that stimulates growth of all hematopoietic lineages. IL-3 is capable of stimulating the pluripotent stem cell, which has recently been isolated in humans. In vivo IL-3 is released by activated T lymphocytes and is probably the least restricted CSF in terms of cell lineage. IL-3 also appears to affect the function of mature granulocytes and monocytes. Clinical trials are currently in progress using recombinant IL-3 (rIL-3) as single-agent therapy and in combination with other CSFs.

Interleukin-4 (IL-4)

First known as B cell stimulatory factor (BSF-1), IL-4 has a number of biological effects. Produced by activated T cells, this agent has the following actions:

- Stimulates growth of resting B cells in vitro
- Increases production of immunoglobulin in vitro
- May stimulate certain T cell lines in vitro
- CSF-like activity in vitro
- May stimulate growth/maturation of mast cells in vitro.

Recombinant IL-4 (rIL-4) is now available, and early phase 1 trials in patients are ongoing.

Erythropoietin

Erythropoietin (Epo) is a hormone produced naturally by the kidneys in response to decreased oxygen levels. Epo stimulates the mitotic activity of erythroid progenitor cells and early precursor cells in the bone marrow, thereby increasing red blood cell production. Epo is now produced in recombinant form in cultured mammalian cells. The initial clinical trials were done evaluating erythropoietin as therapy for chronic anemia in patients with end-stage renal disease. Therapy was given intravenously three times a week, and virtually all patients achieved normalization of hematocrit (HCT) and were transfusion independent. This led to the approval of epoetin alfa (recombinant human erythropoietin) by the FDA for the treatment of anemia in chronic renal failure. Epoetin alfa is generally started at 50–100 U/kg either intravenously or subcutaneously three times weekly. When the desired HCT is achieved, the maintenance dose is individually titrated to maintain that level. In general, epoetin alfa is extremely well tolerated, with adverse events reported as similar to those seen with chronic renal failure. Blood pressure should be adequately controlled before initiation of therapy, and iron supplements may be necessary to maintain iron stores. Clinical trials continue evaluating the efficacy of epoetin alfa for the treatment of anemia in patients with cancer and other disease.

Summary

The addition of new biological agents into clinical trials continues at a rapid rate. Early on, confusion existed because agents were often named for the biological function they stimulated. This method proved problematic, as many agents perform multiple functions in vivo. Now, as agents are identified and the DNA sequence determined, they are given an interleukin designation in numerical order. Currently, interleukin-12 has been identified. Over the next 10 years one can expect numerous clinical trials evaluating both new and previously identified BRMs used alone and in combination with other agents.

PART III

Nursing Management of the Patient Receiving Biological Response Modifiers

Prepared by Kimberly A. Rumsey, RN, MSN, OCN; Paula Trahan Rieger, RN, MSN, OCN; and Margaret Harle, RN, OCN

The professional nurse has a dual role when administering biological response modifiers. The nurse acts as both patient educator and patient advocate, spending a large proportion of her time teaching patients and their significant others about the agents and protocols and training them in self-administration techniques. As patient advocate, the nurse supports the patient's decision, even if the decision might be in conflict with the nurse's set of values. In addition, the nurse plays a major role in identification, documentation, and management of side effects.

After a brief review of the drug development process, Part III will discuss the nursing management of the patient receiving biological response modifiers. The roles of the nurse as patient educator and patient advocate are also further defined. Suggested plans of care for the patient experiencing side effects from BRMs and a chart with the most frequently seen side effects are also included.

The Drug Development Process

The drug development process consists of a preclinical phase and four clinical trial phases. The process begins when a substance with either potential or observed antitumor effects is identified. The substance then goes through a screening process to evaluate its ability to prevent growth of cancer or to cause tumor regression. The second step in preclinical trials, purification and formation, includes refining of the substance, identifying the chemical structure, and getting the substance into a form that can be administered through traditional routes. The compound is then tested in animals in order to establish a safe starting dose that does not produce irreversible toxicities and to identify acute and chronic side effects. After the preclinical trial phase, an application is made to the Food and Drug Administration (FDA) for approval to use the substance in patients with widespread cancer. Goals and eligibility requirements for each of the four clinical trial phases are shown in table 5.

Table 5. Elements of Clinical Trials

	Goals	**Eligibility**
Phase 1	▪ Determine maximum tolerated dose (MTD); highest dose with acceptable toxicity ▪ Identify toxicities and treatment of the toxicities ▪ Determine optimal dose and schedule for future clinical trials	▪ Advanced metastatic cancer ▪ Tumors resistant to standard therapy ▪ Little or no chance of benefiting from standard treatments
Phase 2	▪ Identify antitumor effect in various cancers ▪ Determine response rate in patients with a specific type or stage of disease ▪ Continue to identify toxicities	▪ Disease resistant to standard treatment ▪ Disease must be measurable
Phases 3–4	▪ Comparison against conventional therapy ▪ Identify therapeutic value	▪ May be newly diagnosed, previously untreated patients ▪ Measurable disease

The Nurse as Patient Educator

One of the major roles of the nurse is that of patient educator. The nurse should assess the understanding of the patient and significant other regarding the proposed treatment plan. They should be able to discuss the goal of treatment (diagnostic, therapeutic, or supportive), the treatment regimen (how, where, and when), and the purpose and frequency of laboratory and diagnostic tests. In addition, patients and significant others must also be aware of special requirements such as inpatient admissions and scheduling of outpatient visits and tests. In the case of investigational trials, the nurse should also discuss the patient's understanding of the benefits and risks of the proposed treatment, alternative treatment options, and how long the patient must stay in the vicinity of the study center. Any misunderstandings should be conveyed to the physician.

Patients and significant others may also have financial concerns. Reimbursement for treatment with biological response modifiers can be a problem for patients. Reimbursement is routinely questioned for non-FDA-approved agents and unlabeled indications. It is imperative that the nurse be familiar with reimbursement issues and resources available to assist the patient.

It is the responsibility of the nurse to teach the patient the expected side effects of the drugs administered and necessary self-care measures. Unlike the case with chemotherapy, the etiology and pathophysiology of many of the side effects are unknown. Therefore, appropriate interventions are sometimes difficult to ascertain. Side effects will be reviewed later in this section. The patient must know signs and symptoms that must be reported immediately. Problems that should be reported include: a continuing fever of 40°C that is uncontrolled with acetaminophen or nonsteroidal anti-inflammatory drugs (NSAIDs), weight gain of 5–10 kg over one week, weight loss of 5–10 kg over several weeks, shortness of breath, dizziness, fatigue that leaves one unable to perform activities of daily living, and severe mental status changes.

With some patients, BRM self-administration is possible by subcutaneous or intramuscular injection or via a small-volume intravenous infusion pump. A program of instruction needs to be developed for each individual patient according to his or her educational needs. Patients should be taught basic steps involved in self-administration of injections, including aseptic technique, reconstitution of the drug, drawing up the proper dose, site selection and rotation, administration of the injection, storage of the drug, and disposal of the equipment at home. If patients are using a small-volume intravenous infusion pump, they need to know how to operate the pump, including changing administration setup, troubleshooting the pump, and discontinuing the infusion.

Numerous resources are available to assist the nurse with patient education in biotherapy. Literature is available through the manufacturer for FDA-approved agents, and large cancer centers may have literature available for purchase. The National Cancer Institute also has many publications that are available free of charge.

The Nurse as Patient Advocate

Patients who undergo biotherapy may have numerous psychosocial difficulties that the nurse may be able to help resolve. These patients frequently have exhausted all known conventional therapies for their disease; biotherapy offers them another glimmer of hope. The nurse can assist patients in sorting out alternatives and in making decisions based on their values and goals. The nurse should support decisions that are made by the patient even if the decision is in conflict with the nurse's own values and goals. Treatment failure can be devastating for these patients. The nurse should assure the patient that something can be done (even if it is only that symptoms will be treated) and that they will not be abandoned.

Often, patients are able to communicate more freely with their nurses than with any other member of the health care team. The nurse should address any misunderstandings, unrealistic expectations, and personal feelings (anger, frustration, loss of control) about the treatment. The nurse should convey these concerns and how the issues were handled to the physician and the health care team. In this way, the patient will be provided with consistent information and support. The nurse can make appropriate referrals to help resolve these problems and act as a communication link between the patient and the health care team.

Management of Side Effects

Several side effects of biological response modifiers are common to many of the agents; in addition, each agent has its own unique side effects. The biological agents and their associated side effects are shown in table 6; plans of care for most frequently seen side effects appear in table 7. (Both tables appear at the end of this section.) Side effects may be either acute or chronic. With many agents, when therapy is started an acute side effect profile is seen (e.g., with IFN, constitutional symptoms). These side effects generally decrease with time. However, more chronic side effects will begin to emerge as therapy progresses over time (e.g., with IFN, fatigue, anorexia, mental status changes). Discussing the timing of side effects with patients will help give them a more realistic picture of what to expect.

When patients are receiving treatment with a BRM—especially in a clinical trial—thorough assessment and documentation of side effects becomes imperative. The nurse is in a unique position to observe the patient's response and to help the patient resolve problems as they occur.

Among the most common side effects of biotherapy are the constitutional symptoms that occur with many of the agents. These "flu-like" symptoms are indeed very similar to the chills followed by fever observed during an episode of influenza. Patients should be told to expect chills but that they will pass quickly. In the chills phase, patients should be kept as warm as possible and may need to be medicated with intravenous meperidine. Fevers can be controlled with acetaminophen and/or NSAIDs. Occasionally, these medications may need to be scheduled around the clock. If fever is prolonged or uncontrolled with acetaminophen and NSAIDs, the physician should be notified. Tepid sponge baths, ice packs, and hypothermia blankets will also help to reduce the fever. Acetaminophen and NSAIDs can alleviate headache, myalgias, and arthralgias associated with constitutional symptoms. If the patient is participating in an investigational trial, it is wise to check medications that may be excluded. For example, many protocols specify that NSAIDs may not be given.

Cardiovascular and pulmonary side effects are common with some agents, particularly IL-2, due to the capillary leak syndrome. These symptoms are reversible and are dose-related. Assessment of BUN and creatinine levels, daily weight determination, and strict intake and output measurement can be important in helping to evaluate fluid imbalances. Because patients may experience orthostatic blood pressure changes, they should be taught to rise from a lying position slowly. They also should be taught to report any shortness of breath or dizziness. Administration of fluids, diuretics, and low-dose vasopressors may be indicated.

Among the most alarming symptoms for patients and their significant others are the neurological effects that may occur with some of the agents. These symptoms are also reversible and dose-related. The changes, which may be very subtle, often are first recognized by family members. Central nervous system toxicity may involve minor changes (slowed thinking, decreased concentration, or memory loss) or major problems (somnolence, disorientation, confusion). Patients and significant others should be instructed to report any such changes.

Hematologic side effects can occur, although they tend to reverse more quickly than with chemotherapy and radiation therapy. The nurse should assess the complete blood count, including differential and platelet counts, and, if ordered, the coagulation profile. The patient should be taught reportable signs and symptoms and the self-care measures to implement should counts fall. The nurse should also monitor electrolytes ($K+$, $Mg+2$, $Ca+2$) and liver function tests (SGOT, SGPT, LDH) for any changes. The nurse should report any alterations to the physician and plan appropriate nursing care based on those lab values.

Summary and Outlook

Caring for patients receiving biological response modifiers provides a challenge for the oncology nurse. As patient educator, the nurse must assess the patient's understanding of biotherapy, including the proposed treatment plan, self-care measures, and self-administration techniques. In addition, the nurse should help patients get through the decision-making process and act as their advocate to the health care team. The nurse is also responsible for assessing and managing side effects of the treatment. Although the pathophysiology for many of these side effects remains unknown, there are nursing interventions that can be implemented to improve the quality of life for the patient. Although it is demanding, the role of the nurse in caring for patients receiving biotherapy is also extremely satisfying.

As the knowledge base in immunology continues to expand, and as more agents are approved by the FDA, biotherapy will be used increasingly as a treatment modality. Biological response modifiers will be used in combination with each other as well as with antineoplastic agents, thus increasing the complexity of nursing care for these patients. As use of combination therapy increases, and the complexity of managing side effects grows accordingly, more nurses will be needed with expertise in administration of BRMs and nursing management of patients receiving these agents. In the future, nurses will be challenged to conduct nursing research in order to find new and innovative interventions for managing the care of these patients.

Table 6. Side Effects of Biological Response Modifiers

Agent	Alteration in hematological lab values	Alteration in mental status	Anaphylaxis	Anorexia	Bone pain	Bronchospasm	Capillary leak syndrome	Chills	Desquamation	Diarrhea	Edema, peripheral	Edema, pulmonary	Fever	Fluid retention	Flushing	Headache	Hives	Hypotension	Liver enzymes	Mucositis	Myalgias	Nausea	Pruritis	Rash	Tachycardia	Weight loss	Weight gain	Other side effects
Interferon alpha and beta	+	O	R	+	O	R	R	+	R	O	R	R	+	R	O	+	R	O	+	R	+	O	O	O	O	+	R	Fever dissipates after first week
Interferon gamma	+	O	R	+	O	R	R	+	R	O	R	R	+	R	O	+	R	+	+	R	+	O	O	O	O	+	R	Fever higher and more persistent
GM-CSF**	+	O	R	O	O	R	O	O*	R	O	R	R	+	R	O	O	R	O†	R	R	O	O	O	O	O	O	O	Erythema at injection site
G-CSF**	+	R	R	R	O	R	R	R	R	R	R	R	R	R	O	R	R	R	O	R	R	R	R	R	R	R	R	
Monoclonal antibodies	O	R	O	O	R	O	R	O	R	R	R	R	O	R	O	O	O	O	R	R	O	R	O	O	O	R	R	Side effects depend on what is attached
Tumor necrosis factor	+	O	R	+	R	R	R	+	R	O	R	R	+	R	R	+	R	O	O	R	+	O	R	R	O	+	R	Severe rigors
Interleukin-2	+	O	R	+	R	R	+	+	O	+	+	O	+	+	+	+	R	+	+	O	+	+	+	+	+	+	+	Weight gain during treatment and weight loss (occurs over time due to decrease in appetite)

*Dose-dependent; as dose increases, chills are more regularly seen preceding fever.

**GM-CSF = Granulocyte-macrophage colony-stimulating factor; G-CSF = granulocyte colony-stimulating factor

†Patients may exhibit 10–20 mm decreases in systolic blood pressure; however, symptomatic hypotension is generally seen at higher doses given intravenously.

+ = Common
O = Occasional
R = Rare

Table 7. Plans of Care for Patients Receiving Biological Response Modifiers

Nursing diagnosis (potential/actual)	Outcome	Interventions		
		Assessment	**Therapeutic**	**Educational**
Knowledge deficit related to treatment with biological response modifiers	▪ Patient/significant other will verbalize expected side effects of agent and reportable signs and symptoms ▪ Patient/significant other will demonstrate self-administration of agent if indicated	▪ Assess knowledge of side effects ▪ Assess ability of patient/significant other to be responsible for self-administration		▪ Instruct patient/significant other of expected side effects and reportable signs and symptoms ▪ Instruct patient/significant other in reconstitution of agent ▪ Instruct patient/significant other in I.M. or SQ injection technique
Altered comfort related to:				
—Chills	▪ Patient/significant other will demonstrate self-care behaviors for management of chills ▪ Patient reports acceptable control of chills	▪ Assess severity and duration of chills and document ▪ Monitor vital signs frequently and report changes to physician	▪ Keep patient warm ▪ Avoid icy fluids ▪ Medicate with meperidine 25 mg–50 mg I.V.P.B. if ordered	▪ Instruct patient/significant other in self-management of chills

Table 7. Plans of Care for Patients Receiving Biological Response Modifiers (cont.)

Nursing diagnosis (potential/actual)	Outcome	Interventions		
		Assessment	Therapeutic	Educational
—Fever 101°–104°F (38.4°–40.0°C.)	▪ Patient/significant other will demonstrate self-care behaviors for management of fever ▪ Patient reports acceptable control of temperature	▪ Monitor temperature frequently every 1–4 hours ▪ Report to physician fevers >104° despite antipyretics or other control measures	▪ Encourage p.o. fluids ▪ Medicate with antipyretic as ordered ▪ For temps >103° despite antipyretic, consider —tepid sponge bath or shower —ice packs to temperature-control areas —hypothermia blanket, if indicated	▪ Instruct patient/significant other to monitor temperature ▪ As therapy continues, teach family to monitor temperature pattern in response to medication. Report unusual spikes to health care team. ▪ Instruct patient/significant other in self-management of fever ▪ Instruct patient/significant other to report to physician if fever does not resolve after therapeutic measures

Table 7. Plans of Care for Patients Receiving Biological Response Modifiers (cont.)

Nursing diagnosis (potential/actual)	Outcome	Interventions		
		Assessment	**Therapeutic**	**Educational**
—Pruritis	▪ Patient identifies/ demonstrates measures to control pruritus ▪ Patient reports acceptable control of pruritus	▪ Assess pruritis —onset/duration —characteristics —factors that relieve/ aggravate the condition —treatment used previously	▪ Water-based lotion and/or cream to affected area p.r.n. ▪ Use colloidal oatmeal baths in bathwater (tepid water instead of hot) ▪ Consider use of room humidifier ▪ Administer medications as ordered; may need aggressive around-the-clock administration ▪ Encourage use of distraction (relaxation techniques, reading, needlework, etc.)	▪ Teach signs/ symptoms of skin changes to report ▪ Discuss measures to promote skin hydration ▪ Teach measures to prevent further skin irritation ▪ Teach relaxation techniques as necessary ▪ Teach patient to substitute rub, pressure, or vibration for scratching (e.g., rub arms with lotion)
Pain related to myalgias, arthralgias, headaches	▪ Patient/significant other will demonstrate self-care behaviors for the management of myalgias and headaches ▪ Patient reports acceptable control of pain	▪ Assess need for analgesia	▪ Acetaminophen 650 mg p.o. q. 3–4 hours as ordered ▪ Apply moist heat to aching areas	▪ Instruct patient/ significant other in self-management of myalgias and headaches

Table 7. Plans of Care for Patients Receiving Biological Response Modifiers (cont.)

Nursing diagnosis (potential/actual)	Outcome	Interventions		
		Assessment	Therapeutic	Educational
Altered nutritional status, less than body requirements related to anorexia, mucositis, nausea, vomiting	▪ Patient will maintain adequate nutritional status ▪ Patient/significant other will verbalize understanding of nutritional needs ▪ Patient/significant other will demonstrate measures to increase nutritional intake	▪ Weigh patient regularly and record ▪ Initiate a calorie count if indicated to assess intake	▪ Provide appetizing food consistent with patient preference ▪ Provide high-calorie, low-bulk foods ▪ Use nutritional supplements ▪ Refer to dietitian for consultation ▪ Try small, frequent meals ▪ Try cold foods if patient is sensitive to food odors ▪ Give antiemetics as ordered. In some cases, around-the-clock administration may be necessary ▪ Discuss with health care team the need for tube feedings, I.V.H. if nutritional status becomes severely compromised	▪ Instruct patient/significant other in measures to provide adequate nutrition

Table 7. Plans of Care for Patients Receiving Biological Response Modifiers (cont.)

Nursing diagnosis (potential/actual)	Outcome	Interventions: Assessment	Interventions: Therapeutic	Interventions: Educational
Altered skin integrity related to treatment with biological agents	■ Patient/significant other will verbalize understanding of factors that may cause alteration in skin integrity ■ Skin integrity will be maintained ■ Patient/significant other will demonstrate behaviors to maintain skin integrity	■ Observe skin daily for any breaks, discoloration, redness ■ Observe injection sites for any changes ■ Observe I.V. sites for phlebitis	■ Turn bedfast patient every 2 hours ■ Rotate injection sites for I.M./SQ agents ■ Apply moist heat to any inflamed areas ■ Apply water-based lotion and/or cream to entire body 2–3 times a day for dry desquamation ■ Consider use of room humidifier for dryness	■ Instruct patient/significant other in individual measures for maintenance of skin integrity ■ Instruct patient/significant other in causative factors
Altered thought processes related to treatment with biologicals (e.g., confusion, slowed mentation, somnolence)	■ Patient/significant other will verbalize understanding of factors that may cause alteration in thought processes ■ Patient/significant other will recognize changes in thinking or behavior ■ Patient will maintain reality orientation	■ Assess mental status prior to therapy ■ Assess mental status each shift or clinic visit and when indicated ■ Assess amount of pre-medication and/or analgesics patient is receiving	■ Orient to time and place if necessary ■ Allow verbalization of feelings ■ Refer to mental health professional if indicated ■ Report changes in mental status to physician ■ Initiate safety measures if indicated	■ Instruct patient/significant other how biological response modifier may cause alteration in thought process ■ Inform patient/significant other that this side effect is temporary

Table 7. Plans of Care for Patients Receiving Biological Response Modifiers (cont.)

Nursing diagnosis (potential/actual)	Outcome	Interventions		
		Assessment	Therapeutic	Educational
Fatigue related to treatment with biologicals	▪ Patient maintains independence in activities of daily living and/or uses measures to prevent further immobility	▪ Assess effect of fatigue on activities of daily living ▪ Assess usual pattern of sleep and rest ▪ Assess for signs and symptoms associated with fatigue ▪ Assess for cofactors that may influence fatigue (anemia, poor nutritional intake, etc.)	▪ Encourage patient to arrange most strenuous activities according to peak energy level ▪ Encourage patient to seek assistance with activities of daily living as necessary ▪ Encourage patient to maintain mobility —ambulatory patient: mild exercise; e.g., walking as tolerated —bedridden patient: turn every 2 hours; active/passive range of motion ▪ Encourage patient to take short naps as required to increase energy levels ▪ Treat cofactors that may contribute to fatigue as appropriate	▪ Teach patient/significant other about expected side effect of biotherapy and self-management measures to prevent sequelae

Table 7. Plans of Care for Patients Receiving Biological Response Modifiers (cont.)

Nursing diagnosis (potential/actual)	Outcome	Interventions		
		Assessment	Therapeutic	Educational
Altered oral mucous membranes related to treatment with biologicals	▪ Patient recognizes and reports changes in mucous membranes ▪ Patient demonstrates knowledge of oral hygiene protocol	▪ Perform oral exam to include: —palate —gingiva —dorsum of tongue —undersurface of tongue —floor of mouth —buccal mucosa —oral pharynx —inner surface of lips ▪ Assess normal oral hygiene routine	▪ Encourage patient to perform oral hygiene after meals and at bedtime ▪ Begin use of salt and soda mouth rinses every 2 hours and p.r.n. ▪ Provide lubricant for lips ▪ Encourage use of soft tooth brush ▪ Consider oral irrigations p.r.n. to hydrate mouth and for comfort ▪ Avoid use of agents that further dry the oral mucosa	▪ Explain rationale for prophylactic oral hygiene protocol ▪ Teach proper oral hygiene protocol ▪ Teach signs and symptoms to report to nurse/physician
Diarrhea related to treatment with biological agents	▪ Patient/significant other identifies measures to correct or control diarrhea	▪ Abdominal assessment every 8 hours and p.r.n. ▪ Monitor intake and output every 8 hours ▪ Monitor stools for frequency, volume, consistency Document ▪ Assess perineal/perianal region for skin status	▪ Administer I.V. fluid as ordered ▪ Encourage adequate p.o. fluid intake ▪ Administer medications that control diarrhea as ordered ▪ Encourage hygiene measures. May consider use of sitz bath and/or barrier creams	▪ Teach complications of diarrhea (e.g., fluid electrolyte imbalance, skin breakdown) and prevention measures ▪ Teach signs and symptoms to report

Table 7. Plans of Care for Patients Receiving Biological Response Modifiers (cont.)

Nursing diagnosis (potential/actual)	Outcome	Interventions		
		Assessment	Therapeutic	Educational
Potential for injury related to allergic reaction	▪ Patient will immediately report signs and symptoms of anaphylaxis	▪ Prior to administration, review prior allergic episodes to include —agent thought to be responsible —description of the episode (reaction to agent) ▪ During administration, assess for —shortness of breath, wheezing —sneezing, coughing —local or generalized urticaria and/or erythema, itching —hypotension —tachycardia —cyanosis —unconsciousness —emesis and/or diarrhea	▪ Administer test dose if ordered ▪ Ensure that emergency equipment is available ▪ Ensure that emergency drugs (epinephrine, diphenhydramine, hydrocortisone) are available	▪ Teach patient to report: —pain or tightness in chest; dyspnea —inability to speak —generalized itching —symptoms of uneasiness, agitation, warmth, dizziness —desire to urinate or defecate

Table 7. Plans of Care for Patients Receiving Biological Response Modifiers (cont.)

Nursing diagnosis (potential/actual)	Outcome	Interventions		
		Assessment	Therapeutic	Educational
Altered tissue perfusion related to treatment with biological agents	▪ Patient will maintain adequate blood pressure	▪ Assess —orthostatic vital signs every 1–4 hours —daily weights —intake and output every 8 hours ▪ Assess respiratory and cardiovascular system every 8 hours or clinic visit for —increased heart rate —decreased blood pressure —complaints of dizziness —complaints of shortness of breath —rales ▪ Assess for fluid shifts as evidenced by —peripheral edema —ascites —rales —weight gain —decreased urine output	▪ Administer I.V. fluids, colloids, and vasopressor as ordered ▪ Institute comfort measures as necessary ▪ Institute safety concerns: —ambulate with assistance —rise from lying to sitting position slowly ▪ Encourage p.o. fluid intake	▪ Teach patient/significant other about expected side effect of biotherapy and temporary nature of side effect ▪ Teach patient/significant other to monitor for side effects while at home (daily weight, temperature t.i.d., etc.) and notify physician of any changes or problems ▪ Teach patient/significant other to recognize these signs and symptoms: —peripheral edema —shortness of breath —decreased urinary output —weight gain ≥5 pounds ▪ Teach patient to rise from lying position slowly

Table 7. Plans of Care for Patients Receiving Biological Response Modifiers (cont.)

Nursing diagnosis (potential/actual)	Outcome	Interventions		
		Assessment	**Therapeutic**	**Educational**
Altered coping related to changes in disease status and/or new therapy	▪ Patient recognizes potential/actual stressors and facilitates own coping strategies	▪ Assess patient's perception of stressors and beliefs about the causes ▪ Assess patient's past use of coping strategies ▪ Evaluate patient's ability to problem solve	▪ Encourage patient to verbalize thoughts and feelings about changes in disease status and new therapy ▪ Identify available resources for support and encourage participation (e.g., American Cancer Society, support groups, chaplain, etc.)	▪ Teach relaxation techniques ▪ Teach problem-solving techniques.

APPENDIX A

Annotated Bibliography

Annotated Bibliography

Biotherapy

Abernathy E. Biotherapy: an introductory overview. Oncol Nurs Forum (Suppl) 1987; 14(6): 13–15.
Basic article that introduces terminology and biological agents and approaches being implemented. Includes a glossary of terms.

Dawson MM. Lymphokines and interleukins. Boca Raton: CRC Press, 1991.
Provides an indepth review of the immune response, production of lymphokines, and characteristics of the major biological agents being studied to date. Clinical applications are discussed.

DeVita VJ, Hellman S, Rosenberg SA, eds. Biologic therapy of cancer. Philadelphia: Lippincott, 1991.
A comprehensive overview of the field of biotherapy, including chapters from authors who have been pioneers in the field. Includes biology, preclinical studies, clinical studies, and applications for all major agents.

Dinarello CA, Mier JW. Lymphokines. N Engl J Med 1987; 317(15): 940–45.
Lymphokines and their role in disease and the immune response are discussed in this article.

Foon KA. Biological response modifiers: the new immunotherapy. Cancer Res 1989; 49: 1621–39.
An overview of all biological agents with heavy emphasis on clinical application of monoclonal antibodies.

Foon KA. Advances in immunotherapy of cancer: monoclonal antibodies and interferon. Semin Oncol Nurs 1988; 4(2): 112–19.
An explanation of the actions and indications of interferon and labeled and unlabeled monoclonal antibodies. Also includes future directions for these agents.

Foon KA. Biotherapy of cancer with interleukin-2, colony-stimulating factors, and monoclonal antibodies. Oncol Nurs Forum (Suppl) 1988; 15(6): 13–22.
Description of interleukins, colony-stimulating factors, and interferons. Includes clear, helpful diagrams.

Oldham RK, ed. Principles of cancer biotherapy. New York: Raven Press, 1987.
The first comprehensive textbook on biotherapy. Although not a nursing text, this book is an excellent resource for those specializing in the field of biotherapy. Provides a historical perspective on the field and a thorough discussion of all agents being tested.

Przepiorka D, Rosenthal SN, Bitran JD, eds. The biological therapy of cancer: a clinical symposium. Semin Oncol (Suppl 5) 1988; 15(5): 1–57.
A symposium focusing on advances in the use of biological proteins in human cancers, with a focus on applications of interferons and colony-stimulating factors. Reimbursement issues are also covered.

Rosenberg SA, Longo DL, Lotze MT. Principles & applications of biologic therapy. In: DeVita VT Jr, Hellman S, Rosenberg SA, eds. Cancer: principles & practice of oncology. 3rd ed. Philadelphia: Lippincott, 1989; 301–47.
Although written for the physician, this chapter provides an in-depth overview of the subject of biotherapy.

Biological actions, clinical applications, supporting clinical studies, and expected toxicities are all covered. Excellent tables and 465 references.

Colony-Stimulating Factors

Clark SC, Kamen R. The human hematopoietic colony-stimulating factors. Science 1987; 236: 1229–37.
Discusses the basic science behind the discovery of the four classic CSFs and their physiology. Although technical, it has a good diagram of hematopoiesis interpolated with the action sites of the four major CSFs.

Gabrilove JL. Colony-stimulating factors: clinical status. In: DeVita VT Jr, Hellman S, Rosenberg SA, eds. Important advances in oncology. Philadelphia: Lippincott, 1991; 215–37.
A comprehensive overview of the physiology of the four major CSFs and a review of major clinical trials investigating their use.

Gabrilove JL, ed. Clinical applications of hematopoietic growth factors. Semin Hematol (Suppl 2) 1989; 26(2): 1–23.
An entire issue exploring potential clinical applications of the CSFs. An overview of physiology is provided, as well as a review of the major studies conducted using erythropoietin, GM-CSF, G-CSF, and CSFs after bone marrow transplant.

Groopman JE. Clinical applications of colony-stimulating factors. Semin Oncol (Suppl 6) 1988; 15(6): 27–33.
A review article that focuses mainly on granulocyte-macrophage colony-stimulating factor (GM-CSF) in patients with AIDS, myelodysplastic syndrome, and cancer following intensive chemotherapy.

Haeuber D, Spross J. Alterations in protective mechanisms: hematopoiesis and bone marrow depression. In: Baird

SB, McCorckle R, Grant M, eds. Cancer nursing: a comprehensive textbook. Philadelphia: Saunders, 1991; 759-81.
Written from a nursing perspective, provides an overview of the process of hematopoiesis and how it relates to the care of cancer patients.

Haeuber D, DiJulio JE. Hemopoietic colony-stimulating factors: an overview. Oncol Nurs Forum 1989; 16(2): 247-55.
An excellent overview of the CSFs, their potential clinical uses, and nursing implications.

Mertelsmann R, Herrmann F, eds. Hematopoietic growth factors in clinical applications. New York: Marcel Dekker, 1990.
An excellent overview of the CSFs being utilized/studied to date. Provides background on biology as well as clinical applications. Good reference for those desiring a more in-depth knowledge base.

Metcalf D. The colony-stimulating factors: discovery, development and clinical applications. Cancer 1990; 65(10): 2185-95.
An excellent overview, written by one of the pioneers in the field, that provides a historical perspective and current clinical applications of the CSFs.

Zanjani ED, Ascensao JL. Erythropoietin. Transfusion 1989; 29(1): 46-57.
An overview of the development and clinical applications of erythropoietin.

Interferon

Figlin RA. Biotherapy in clinical practice. Semin Hematol (Suppl 3) 1989; 26(3): 15-24.
An overview of the use of interferon for the treatment of specific diseases. Provides a historical perspective of the

development of interferon and discusses controversial issues such as dosing and formation of antibodies.

Gutterman JU. The role of interferons in the treatment of hematologic malignancies. Semin Hematol (Suppl) 1988; 25(3): 3–8.
Discusses the clinical applications of interferons in the treatment of hematologic malignancies.

Horoszewicz JS, Murphy GP. Interferon: A review of its development, clinical use and future applications. Hosp Formulary 1987; 22(9): 776–96.
Provides an excellent overview of the biological effects of interferon with good diagrams, and also covers early clinical trials, toxicities, and potential applications.

Kirkwood JM, Ernstoff MS. A clinical update: The role of interferon in the biotherapy of solid tumors. Oncol Nurs Forum (Suppl) 1988; 15(6): 3–6.
Short article discussing the role of alpha and gamma interferon in the treatment of solid tumors. In addition, the need to establish an optimal biological response modifying dose is summarized.

Kirkwood JM, Ernstoff MS. Interferons in the treatment of human cancer. J Clin Oncol 1984; 2(4): 336–52.
One of the classic review articles discussing the biology and clinical application of interferon.

Mandelli F, Avvisati G, Amadori S, et al. Maintenance treatment with recombinant interferon alfa-2b in patients with multiple myeloma responding to conventional induction chemotherapy. N Engl J Med 1990; 322(20): 1430–34.
A pivotal study looking at the use of alfa interferon as adjuvant therapy in patients with multiple myeloma.

Pazdur R, Jackson D, Shepherd B, et al. 5-fluorouracil and recombinant interferon alfa-2a: review of activity and toxicity in advanced colorectal carcinomas. Oncol Nurs Forum (Suppl) 1991; 18(1): 11–18.
A concise overview of studies evaluating the combina-

tion of alfa interferon and 5-FU. Provides information on potential mechanisms of action for this combination and provides a table for grading of toxicity. Three case studies are presented.

Pestka S. The purification and manufacture of human interferons. Sci Am 1983; 249(2): 36–43.
Article provides a historical perspective on the process of how interferon came to be produced by recombinant DNA technology. Excellent diagrams.

Quesada JR, Talpaz M, Rios A, Kurzrock R, Gutterman JU. Clinical toxicity of interferons in cancer patients: a review. J Clin Oncol 1986; 4(2): 234–43.
The classic article reviewing the toxicity of interferon. A systems approach is used to outline the side effects.

Interleukin-2

Gambacorti-Passerini C, Sondell PM. Problems, controversies, and perspectives on the clinical use of interleukin-2. Curr Opinion Oncol 1990; 2: 1139–45.
Focuses on the analysis and the possible circumvention of the distinct problems that presently appear to be preventing the clinical development of interleukin-2 therapy as an effective standard cancer treatment.

Parkinson DR. Interleukin-2 in cancer therapy. Semin Oncol (Suppl) 1988; 15(6): 10–26.
Provides an indepth overview of the biology of IL-2 and current clinical trials. Also provides a thorough review of side effects by system, and an excellent table of side effects.

Rosenberg SA. Adoptive immunotherapy for cancer. Sci Am 1990; 262(5): 62–69.
A brief overview of the immune system and a historical overview of the development of adoptive immunotherapy. Excellent diagrams and photographs of clinical responses.

Rosenberg SA. Immunotherapy of patients with advanced cancer using interleukin-2 alone or in combination with lymphokine activated killer cells. In: DeVita VT Jr, Hellman S, Rosenberg SA, eds. Important advances in oncology. Philadelphia: Lippincott, 1988; 217–57.
An in-depth review of studies using IL-2 and IL-2 plus LAK in cancer patients. Discusses different regimens, toxicities, and responses.

Rosenberg SA, Lotze MT, Yang JC, et al. Combination therapy with interleukin-2 and alpha interferon for the treatment of patients with advanced cancer. J Clin Oncol 1989; 7(12): 1863–74.
A review of 94 patients treated with this combination. Discusses responses, toxicities, and future directions.

Siegel JP, Puri RK. Interleukin-2 toxicity. J Clin Oncol 1991; 9(4): 694–704.
A comprehensive overview of the toxicities experienced with IL-2 administration. Pathophysiology, when known, is also discussed.

Simpson C, Seipp CA, Rosenberg SA. The current status and future applications of interleukin-2 and adoptive immunotherapy in cancer treatment. Semin Oncol Nurs 1988; 4(2): 132–41.
An excellent and comprehensive overview of clinical trials using IL-2 and effector cells. Good toxicity table for high-dose IL-2.

Smith KA. Interleukin-2. Sci Am 1990; 262(3): 50–57.
An easily read overview of the biology of interleukin-2. Includes excellent diagrams.

Monoclonal Antibodies

DiJulio JE. Treatment of B cell and T cell lymphomas with monoclonal antibodies. Semin Oncol Nurs 1988; 4(2): 102–06.

A limited description of two clinical trials using monoclonal antibodies in the treatment of B cell and T cell lymphomas. A brief summary of nursing care is included.

Dillman JB. New antineoplastic therapies and inherent risks: monoclonal antibodies, biologic response modifiers, and interleukin-2. J Intravenous Nurs 1989; 12(2): 103–113.
An excellent overview of biotherapy with an emphasis on monoclonal antibodies written for the nurse. Biology of the agents are discussed, as well as pertinent clinical studies and nursing management.

Dillman JB. Toxicity of monoclonal antibodies in treatment of cancer. Semin Oncol Nurs 1988; 4(2): 107–111.
A comprehensive review article on toxicities associated with monoclonal antibodies. Discusses toxicities in terms of timing (e.g., acute, delayed) and appropriate nursing actions. Good review for those with a beginning knowledge of this area.

Dillman RO. Monoclonal antibodies for treating cancer. Ann Intern Med 1989; 111: 592–603.
An excellent overview of the use of monoclonal antibodies for the treatment of cancer for those desiring more in-depth knowledge.

Dillman RO, Beauregard JC, Halpern SE, et al. Toxicities and side effects associated with intravenous infusions of monoclonal antibodies. J Biol Response Mod 1986; 5: 73–84.
The classic article reviewing toxicity associated with monoclonal antibody therapy. Includes data from 186 infusions of antibody.

McCullough KC, Spier RE, eds. Monoclonal antibodies in biology and biotechnology: theoretical and practical aspects. Cambridge: Cambridge University Press, 1990.
A current review of production and application of monoclonal antibodies. Includes an overview of the immune system with emphasis on humoral immunity, and hy-

bridoma production. The final chapter discusses applications both clinical and scientific. A reference list by chapters is at the end. This book is mainly for those desiring more technical rather than clinical knowledge.

Milstein C. Monoclonal antibodies. Sci Am 1980; 243(4): 66–74.
An excellent overview of the process by which monoclonal antibodies are made, written by one of the co-discovers of hybridoma technology. Excellent diagrams.

Moldawer NP, Murray JL. The clinical uses of monoclonal antibodies in cancer research. Cancer Nurs 1985; 8(4): 207–13.
One of the first nursing articles written on monoclonal antibody therapy. Written at an understandable level, it covers nursing care as well as major potential applications of monoclonals therapeutically and diagnostically.

Rieger PT. Monoclonal antibodies: target-specific magic bullets. Am J Nurs 1987; 87(4): 469–73.
This article is targeted to the nurse generalist, and discusses what monoclonal antibodies are, how they are made, and how they might be used clinically. A good beginner article for someone who has little to no background in monoclonal antibodies.

Tumor Necrosis Factor

Blick M, Sherwin S, Rosenblum M, et al. Phase I study of recombinant tumor necrosis factor in cancer patients. Cancer Res 1987; 47: 2986–89.
One of the early clinical studies using tumor necrosis factor. Discusses dosing and toxicity.

Frei E, Spriggs D. Tumor necrosis factor: still a promising agent. J Clin Oncol 1989; 7(3): 291–94.
An interesting editorial on the status of TNF. Examines the disappointment in clinical trials and reasons for

why this has occurred. Provides direction for future research.

Moldawer N, Figlin R. Tumor necrosis factor: current clinical status and implications for nursing management. Semin Oncol Nurs 1988; 4(2): 120–25.
This article provides a review of the biology, clinical trials conducted, toxicities experienced, and nursing management required for tumor necrosis factor. Good table summarizing toxicities.

Old LJ. Tumor necrosis factor (TNF). Science 1985; 230: 630–32.
Review of the biology of TNF for those desiring more detailed information. Gives a good historical perspective of the discovery and development of TNF, easily readable.

Silverman P, Berger NA. Recent clinical experience with tumor necrosis factor and advances in understanding its physiologic function and cellular activities. Curr Opinion Oncol 1990(2); 1133–38.
Provides a review of the in vitro work with TNF and its physiologic role in disease states. Examines the question of why TNF has not been successful as a single agent.

Tracey KJ, Vlassara H, Cerami A. Cachectin/tumor necrosis factor. Lancet 1989; 1(8647): 1122–26.
An excellent overview of the biology of TNF for those desiring in-depth information. Discusses TNF's role in septic shock, cachexia, inflammation, tissue remodeling and malignancy.

Immunology

Burnet FM. Immunology, aging, and cancer: medical aspects of mutation and selection. San Francisco: W.H. Freeman, 1976.

This classic text provides an overview of the immune system, and its interaction in disease states. The theory of immune surveillance is presented.

Carpenter CB. The major histocompatibility gene complex. In: Wilson JD, Braumwald E, Isselbache KJ, et al, eds. Harrison's principles of Internal Medicine. 12th ed. New York: McGraw Hill, 1991; 86–92.
A thorough, but technical review of the MHC. Several good diagrams; covers application of genetic defects in HLA to disease states.

Gallucci BB. The immune system and cancer. Oncol Nurs Forum, (Suppl) 1987; 14(6): 3–12.
A review of the components and underlying principles of the immune system. Various aspects of tumor immunology are explored.

Grady, C. Host defense mechanisms: an overview. Semin Oncol Nurs 1988; 4(2): 86–94.
A fundamental overview of host defense mechanisms. Nonspecific and specific immunities are outlined.

Griffin JP. Hematology and immunology: concepts for nursing. Norwalk, CN: Appleton-Century Crofts, 1986.
A textbook written for nurses that discusses the immune system and then relates the information clinically to hematologic malignancies.

Guyton AC. Textbook of medical physiology. 8th ed. Philadelphia: Saunders, 1991; 365–82.
A comprehensive, easily read, well diagrammed textbook of physiology. Excellent unit on blood cells, immunity, and blood clotting.

Haynes BF, Fauci AS. The immune system. In: Wilson JD, Braumwald E, Isselbache KJ, et al, eds. Harrison's principles of internal medicine. 12th ed. New York: McGraw Hill, 1991; 76–86.
Review of the immune system that discusses cells involved in the immune response, their functions, and

interactions with other cells. Excellent diagrams and tables. Also discusses evaluation of immune function.

Jaret P. Our immune system: the wars within. Nat Geographic 1986; 169(6): 702–735.
An excellent overview of the immune system written for the layman. Includes electron micrographs of immune cells at work, as well as good diagrams to illustrate concepts. Uses AIDS as example to illustrate destruction of the immune system.

Jaroff L. Stop that germ. Time, May 23, 1988. 56–64.
An overview of the immune system with different clinical vignettes highlighted. Written for the lay person, includes diagrams.

Nossal GJV. The basic components of the immune system. N Engl J Med 1987; 316(21): 1320–25.
A key article describing basic components of the immune system. Includes: the encounter between antigen and lymphocytes, activation of lymphocytes, deployment of immunologic messages, discrimination between self and nonself, and regulation of the immune responses.

Roitt IM, Brostoff J, Male DK, eds. Immunology. 2nd ed. St. Louis: Mosby, 1989.
This comprehensive text on immunology is best known for its excellent color diagrams that illustrate the material presented. Several chapters also include application of immunologic concepts in clinical situations.

Schindler LW. Understanding the immune system. U.S. Dept Health Hum Serv. Bethesda, MD: Nat Cancer Inst (NIH pub #88–529), 1988.
Although written for patients, this overview of the immune system is more appropriate for health care personnel. Written in a clear, readable manner with excellent diagrams and glossary.

Stites DP, Terr AI, eds. Basic and clinical immunology. 7th ed. Norwalk, CN: Appleton and Lange, 1991.

A comprehensive overview of basic immunology, laboratory tests, and clinical immunology. Although at times complex, a strong point of this text are the numerous chapters that provide a focus on primary immunologic diseases or on disorders with important immunopathologic characteristics.

Psychoneuroimmunology

McHugh MK, Tri Van Nguyen F, eds. Psychosocial nursing and immunocompetence. Holistic Nurs Pract 1991; 5(4): 1–38.
A current review of the concept of psychoneuroimmunology and how it impacts nursing.

Korneva EA, Klimenko VM, Shkhinek EK, eds. Neurohumoral maintenance of immune homeostasis. Chicago: University of Chicago Press, 1985.
A detailed examination of the investigation into the interactions between the nervous system and the immune system. This text is somewhat complex.

Bovbjerg DH. Psychoneuroimmunology: implications for oncology? Cancer (Suppl) 1991; 67(3): 828–32.
Discusses how psychosocial influences on immune function could potentially prove a mechanism to account for reports of association between psychosocial factors and cancer prognosis.

Nursing

Brophy LR, Sharp EJ. Physical symptoms of combination biotherapy: a quality-of-life issue. Oncol Nurs Forum (Suppl) 1991; 18(1): 25–30.
Provides a discussion on the physical symptoms secondary to treatment with alpha interferon and 5-FU and alpha interferon and interleukin-2. Quality of life issues

and nursing interventions to impact the physical symptoms are reviewed.

Carter P, Engelking C, Rumsey K, et al. Biological response modifier guidelines: recommendations for nursing education and practice. Pittsburgh: Oncology Nursing Society, 1989.
Recommendations for biological response modifiers course content and clinical practicum. Includes a detailed list of side effects and their management.

Haeuber D. Recent advances in the management of biotherapy-related side effects: flu-like syndrome. Oncol Nurs Forum (Suppl) 1989; 16(6): 35–47.
Provides an in-depth discussion of symptoms associated with the flu-like syndrome and the pathophysiology behind them. Methods to prevent or alleviate these side effects are also presented. An excellent overview.

Hahn MB, Jassak PF. Nursing management of patients receiving interferon. Semin Oncol Nurs 1988; 4(2): 95–101.
A good summary of interferon, including cellular effects, pharmacokinetics, therapeutic uses, and administration information. Toxicity management according to systems and other nursing concerns such as patient education, handling and disposal, and ethical considerations are examined.

Hogan CM. Coping with biotherapy: physiological and psychosocial concerns. Oncol Nurs Forum (Suppl) 1991; 18(1): 19–23.
A brief overview on helping patients to cope with biotherapy. The article discusses factors unique to biotherapy, assessment of coping patterns, and how adult developmental level may affect the ability to cope.

Hood LE, Abernathy E. Biological response modifiers. In: Baird SB, McCorckle R, Grant M, eds. Cancer nursing: comprehensive textbook. Philadelphia: Saunders, 1991; 321–43.

Irwin MM. Patients receiving biological response modifiers: overview of nursing care. Oncol Nurs Forum (Suppl) 1987; 14(6): 32–37.
Using interferon as a paradigm, this article provides a concise overview of the nursing care required for patients receiving BRMs.

Jassak PF. Biotherapy. In: Groenwald SL, Hansen Frogge M, Goodman M, Yarbro CH, eds. Cancer nursing: principles and practice. Boston: Jones & Bartlett, 1990; 284–306.
This chapter on biotherapy covers definition, historical perspective, an overview of all major agents that includes cellular effects, clinical uses, administration and toxicity. Nursing concerns are discussed comprehensively at the end of the chapter. Well written, 199 references.

Jassak PF, Sticklin LA. Interleukin-2: an overview. Oncol Nurs Forum 1986; 13(6): 17–22.
One of the initial nursing papers published on IL-2. Provides a description of the phase 2 clinical trials using lymphokine activated killer (LAK) cells and IL-2. The role of the research nurse in caring for patients in clinical trials is defined.

Lynch MT. The nurse's role in the biotherapy of cancer: clinical trials and informed consent. Oncol Nurs Forum (Suppl) 1988; 15(6): 23–27.
A brief but thorough overview of the nurse's role in informed consent. Provides a discussion on the ethical/legal issues surrounding informed consent and the nurse's role as educator and advocate.

Lynch MT, Yanes L, Todd K. Nursing care of AIDS patients participating in a phase 1-2 trial of recombinant human granulocyte-macrophage colony-stimulating factor. Oncol Nurs Forum 1988; 15(4): 463–469.
The first nursing article written on colony-stimulating factors (CSFs). Discusses the initial clinical trial in AIDS

patients. Provides an overview of hematopoiesis, protocol overview, and side effects. Nursing implications include a detailed care plan for patients receiving GM-CSF.

Mayer DK. Biotherapy: recent advances and nursing implications. Nurs Clin N Am 1990; 25(2): 291–308.
An up-to-date overview of the field of biotherapy, agents being used clinically, their biological actions, and a complete nursing care plan for side effects experienced.

Morra ME, Grant M, eds. Cancer patient education. Semin Oncol Nurs 1991; 7(2): 79–145.
A complete issue devoted to the subject of patient education. Numerous topics are reviewed, including why patient education is needed, theories and models used to design programs, practical aspects of identifying content, and implementing programs in various care settings. Methods for evaluating programs and materials are also presented.

Padavic-Shaller K. IL-2: nursing applications in a developing science. Semin Oncol Nurs 1988; 4(2): 142–50.
A well written, easily readable article covering the toxicities associated with IL-2 by system. Includes a comprehensive nursing care plan. Excellent reference to guide the clinical care of patients receiving IL-2.

Piper BF, Rieger PT, Brophy L, et al. Recent advances in the management of biotherapy-related side effects: fatigue. Oncol Nurs Forum (Suppl) 1989; 16(6): 27–34.
A comprehensive review of the current knowledge on biotherapy related fatigue to guide nursing practice and provide direction for future nursing and clinical trial research.

Rieger PT. Biotherapy. In: Otto S, ed. Oncology nursing. St. Louis: Mosby Year Book, 1991; 318–48.
This chapter includes a historical perspective on the development of biotherapy, biological actions and clinical applications of all major agents, and side effects.

Excellent overview of nursing implications for patients receiving biotherapy.

Collins JL, Thaney KM, eds. Biotherapy: a nursing challenge. Semin Oncol Nurs 1988; 4(2): 83–150.
An excellent overview on biotherapy with the entire issue dedicated to the subject. Includes an overview on the immune system, articles on each major biological response modifier, and nursing management.

Yasko JM. A progress report on reimbursement of biotherapy: prospects for the 1990s. Oncol Nurs Forum (Suppl) 1989; 16(6): 42–47.
A concise review of trends affecting clinical trial reimbursement, current status, and recent initiatives attempting to combat the problem.

APPENDIX B

Patient Teaching Resources

Patient Teaching Resources

Roche Laboratories. Roferon-A package insert. Nutley, NJ: Hoffmann-LaRoche, 1987.

Schering Corporation. Intron A : interferon alfa-2b recombinant for injection. Kenilworth, NJ: Schering, 1986.

Amgen, Inc. Getting ready for Epogen—Information for our dialysis patients (flip chart). Thousand Oaks, CA: Amgen, 1990.

Amgen, Inc. Self-injection chart: step-by-step guide to subcutaneous self-injection (English and Spanish). Thousand Oaks, CA: Amgen, 1991.

Roche Laboratories. At home with your Roferon-A therapy: patient guide (available in Spanish). Nutley, N.J.: Hoffmann-LaRoche Inc., 1990.

Roche Laboratories. Managing your patients on Roferon-A: guidelines for the health care professional. Nutley, NJ: Hoffmann-LaRoche, 1990.

Roche Laboratories. Roferon-A administration kit. Nutley, N.J.: Hoffmann-LaRoche Inc., 1990.

Roche Laboratories. Roferon-A subcutaneous self-administration instruction review (flip chart). Nutley, NJ: Hoffmann-LaRoche, 1990.

Roche Laboratories. The right combination to increase survival and improve quality of life. Nutley, NJ: Hoffmann-LaRoche, 1990.

Schering Corporation. Flexible dosing chart: Intron-A. Kenilworth, NJ: Schering, 1990.

Schering Corporation. Intron A: a day-by-day guide. Kenilworth, NJ: Schering, 1989.

Schering Corporation. Self administration of Intron A: step-by-step subcutaneous injection technique. Kenilworth, NJ: Schering, 1989.

Schering Corporation. Taking control of your therapy. Kenilworth, NJ: Schering, 1989.

Schering Corporation. Intron A: a practical guide for nurses. Kenilworth, NJ: Schering, 1986.

Schering Corporation. Living with Intron A (kit). Kenilworth NJ: Schering, 1991.

U.S. Department of Health and Human Services. Making health communication programs work: a planner's guide. Bethesda, MD: National Cancer Institute (NIH Publication #89-1493), 1989.

Amgen, Inc. How to give yourself a subcutaneous injection (videotape). Thousand Oaks, CA: Amgen, 1991.

Roche Laboratories. At home with Roferon-A therapy (videotape and audiotape cassette). Nutley, NJ: Hoffmann-LaRoche, 1990.

Roche Laboratories. The right start with Roferon-A—managing side effects (videotape). Nutley, NJ: Hoffmann-LaRoche, 1990.

Roche Laboratories. Roferon-A cost assistance program (videotape). Nutley, NJ: Hoffmann-LaRoche, 1990.

Schering Corporation. Self-Injection of Intron-A (videotape). Kenilworth, NJ: Schering, 1990.

Roche Laboratories. Roferon-A cost assistance program: program guide. Nutley, NJ: Hoffmann-LaRoche, 1989.

Schering Corporation. Schering patient assistance program: IRIS, an innovative approach to reimbursement. Kenilworth, NJ: Schering, 1989.

Hoechst-Roussel Pharmaceuticals, Inc. The Hoechst reimbursement information service. Sommerville NJ: Hoechst-Roussel, 1991.

Amgen, Inc. Neupogen reimbursement hotline. Thousand Oaks, CA: Amgen, 1990.

Immunex Corporation. A reimbursement support program for Leukine. Seattle: Immunex, 1990.

APPENDIX C

Self-Testing Materials

Self-Testing Materials—Questions

PART I. The Immune System

Organs of the Immune System

1. List the three functions of the immune system.

   A.

   B.

   C.

2. The two subdivisions of the immune system are the __________ and the __________ immune systems.

3. Match the following:

   __ A. Thymus
   __ B. Lymph nodes
   __ C. Tonsils
   __ D. Bone marrow
   __ E. Lymph vessels
   __ F. Spleen
   __ G. Peyer's patches

   I. Primary lymphoid organ
   II. Secondary lymphoid organ

4. The major function of the bone marrow is __________.

Cells of the Immune System

1. The basic cell of hematopoiesis is the ______________.

2. The stem cell can develop into either the _________ cell line, which develops into red blood cells and platelets, or the _________ cell line, which develops into T and B lymphocytes.

3. List the three types of granulocytes: _________, _________, _________.

4. The major role of the granulocyte is _________.

5. _________ circulate in the blood for about 12 hours, then migrate to the tissues and die after about 48 hours.

6. _________ generate chemotactic factors that attract cells to the site of inflammation.

7. Macrophages are _________ that have migrated to the tissues.

8. Match the following:

__ A. B lymphocytes	I. Responsible for humoral immunity
__ B. T lymphocytes	II. Responsible for cell-mediated immunity

Innate Immunity

1. The function of the innate immune system is to:

2. Name three mechanical barriers and how they prevent invasion.

   A.

   B.

   C.

3. Describe the process of phagocytosis.

4. List the five cardinal signs of inflammation.

   A.

   B.

   C.

   D.

   E.

5. Describe the physiologic process involved in the inflammatory response.

6. Interferon is a soluble factor that defends the body against __________.

7. List two functions of natural killer cells.

   A.

   B.

8. List complement's three mechanisms of mediating the inflammatory process.

   A.

   B.

   C.

Adaptive Immunity

1. The two hallmarks of the adaptive immune system are __________ and __________.

2. Upon second exposure to an organism, ________________ ensure an accelerated and increased immune response.

3. The two major types of effector mechanisms of adaptive immunity are __________ and ________________.

4. Interaction does occur between the __________ and the __________ immune responses.

5. ________________ function as antibodies.

6. B cells, when stimulated by antigens, differentiate into __________ __________, which secrete antibodies.

7. Match the following:

__ A. IgG	I. Function unknown
__ B. IgM	II. Function in hypersensitivity reaction
__ C. IgA	III. Found in seromucinous secretions
__ D. IgD	IV. Found in large amounts during initial encounter with an antigen
__ E. IgE	V. Functions during the secondary immune response

8. Describe two of the mechanisms by which antigens are eliminated.

A.

B.

9. The __________ antibody response is much faster due to the presence of memory cells.

10. The majority of circulating lymphocytes are __________.

11. __________ on the surface of the cell assist in dividing T lymphocytes into subsets.

12. Match the following:

__ A. Helper T cells	I. Decrease the response of T and B cells
__ B. Suppressor T cells	II. Circulate in a "resting state"
__ C. Cytotoxic T cells	III. Kill cells by lysis
__ D. Memory cells	IV. Help initiate the immune response

13. Define lymphokine.

14. Name three cytokines.

    A.

    B.

    C.

15. The unique antigen code found on the surface of body cells allows the immune system to distinguish between __________ and __________.

16. In organ and bone marrow transplantation, an attempt is made to __________ several alleles between the donor and recipient.

PART II. Biological Response Modifiers

Rationale for Use of Biotherapy

1. Describe the theory of immune survelliance.

2. The rationale for the use of lymphokines, adoptive immunotherapy, and other biological response modifers is based on observations of the interaction between __________ __________ and the __________ __________.

3. List four factors thought to affect immune function.

   A.

   B.

   C.

   D.

4. Psychoneuroimmunology can be defined as the study of the effects of __________, __________ __________, __________ __________, and __________ on the immune system.

Historical Perspectives

1. List two major technological advances that have assisted biotherapy in becoming a fourth modality in the treatment of cancer.

2. Match the following:

   __ A. Biotherapy

   __ B. Biological response modifier

   __ C. Biologicals

   I. Agents or approaches that utilize mechanisms of action which involve the patient's own biological response

   II. Agents derived from biological sources and/or use of agents that affect biological responses

   III. Agents extracted from or produced from biological material

Classification of Agents

1. List the three major classifications of BRM.

   A.

   B.

   C.

Agents

1. List the three types of interferon.

   A.

   B.

   C.

2. List the three major actions of interferon.

   A.

   B.

   C.

3. __________ interferon was the first approved by the Food and Drug Administration.

4. Interleukin-2 is also known as __________ __________ __________ __________.

5. Name two types of effector cells that are currently being studied.

   A.

   B.

6. Describe the proposed mechanism of action of tumor necrosis factor.

7. The technique used for producing monoclonal antibodies is known as ________ ________.

8. Describe a monoclonal antibody.

9. List two potential uses of monoclonal antibodies.

   A.

   B.

10. Describe the physiological role of colony-stimulating factors.

11. List two potential clinical applications of colony-stimulating factors.

    A.

    B.

12. Match the following:

__ A. GMCSF	I. Stimulates growth for all hematopoietic lineages
__ B. M-CSF	II. Stimulates production of neutrophils and monocytes
__ C. IL-3	III. B cell growth factor
__ D. IL-4	IV. Stimulates production of monocyte progenitors

PART III. Nursing Management of the Patient Receiving Biological Response Modifiers

The Drug Development Process

1. Match the following:

__ A. Phase 1	I. To compare with conventional therapy
__ B. Phase 2	II. To determine maximum tolerated dose, dosage, and scheduling
__ C. Phases 3–4	III. To determine response rate for a specific type or stage of disease

2. Match the following:

__ A. Phase 1	I. Disease is resistant to standard treatment and must be measurable
__ B. Phase 2	II. May be previously untreated patients
__ C. Phases 3–4	III. Patient has advanced cancer and little or no chance of benefiting from standard treatment

The Nurse as Patient Educator

1. List the five points concerning the proposed treatment plan that the patient and significant other should be able to discuss.

   A.

   B.

   C.

   D.

   E.

2. List three additional topics for patient teaching.

   A.

   B.

   C.

The Nurse as Patient Advocate

1. The nurse may act as the communication link between the __________ and the __________ __________ __________.

2. The nurse should also __________ decisions that are made by the patient, even if the decision is in conflict with the nurse's set of values.

Self-Testing Materials—Answers

PART I. The Immune System

Organs of the Immune System

1. defense, homeostasis, surveillance

2. innate, adaptive

3.
   A. I
   B. II
   C. II
   D. I
   E. II
   F. II
   G. II

4. Hematopoiesis

Cells of the Immune System

1. pluripotent stem cell

2. myeloid, lymphoid

3. neutrophils, eosinophils, and basophils

4. phagocytosis

5. neutrophils

6. basophils

7. monocytes

8.
   A. I
   B. II

Innate Immunity

1. Provide the body's primary line of defense against foreign invasion

2. Any three:
   A. Epidermis—tough protective barrier
   B. G.I. tract—layer of epithelium, endogenous bacteria, digestive enzymes, motility and gastric acid
   C. Respiratory tract—small hairs, ciliated epithelium, cough reflex
   D. G.U. system—properties of bladder mucosa, flushing action of urine

3. Phagocytic cells engulf invading microorganisms and enclose them in the cytoplasm, thereby preventing their proliferation

4.
   A. Redness
   B. Swelling
   C. Heat
   D. Pain
   E. Loss of function (sometimes)

5. Invasion of the microorganism causes release of chemical mediators, resulting in vasodilation and increased blood flow to the area. Capillary permeability increases, phagocytes migrate to the area, fluid shifts to the interstitial space, and fibrin clots surround the area

6. viruses

7.
   A. Kill virally infected cells
   B. Kill cancerous cells

8.
   A. Lysis
   B. Opsonization
   C. Chemotaxis

Adaptive Immunity

1. specificity and memory

2. memory cells

3. humoral, cell-mediated

4. innate, adaptive

5. immunoglobulins

6. plasma cells

7.
   A. V
   B. IV
   C. III
   D. I
   E. II

8. Any two:
   A. Neutralization—antibody covers up sites on antigen, rendering it harmless
   B. Opsonization—antigen-antibody complex becomes sticky and is more easily phagocytized
   C. Agglutination—clumping together of antigen-bearing cells in the presence of antibodies
   D. Complement—complement-mediated cell lysis

9. secondary

10. T lymphocytes

11. Surface proteins or surface markers

12.
   A. IV
   B. I
   C. III
   D. II

13. A chemical messenger with which cells of the immune system communicate. It is produced by lymphocytes.

14. Any three:
   A. Interleukin-1
   B. Interleukin-2
   C. Interleukin-3
   D. Colony-Stimulating Factors
   E. Interleukin-4
   F. Interleukin-6
   G. Gamma Interferon
   H. Interferon (alpha, beta)
   I. Tumor necrosis factor

15. self, nonself

16. match

PART II. Biological Response Modifiers

Rationale for Use of Biotherapy

1. Tumors are recognized as foreign and destroyed by the immune system

2. malignant cells, immune system

3.
   A. Age
   B. Heredity
   C. Nutritional status (obesity)
   D. Emotional status (stress)

4. behavior, emotional status, social interactions, stress

Historical Perspectives

1. Any two:
   A. Increased understanding of the immune system
   B. Recombinant DNA technology
   C. Hybridoma technology

2.
   A. II
   B. I
   C. III

Classification of Agents

1.
   A. Augment, modulate, or restore the host's immune response
   B. Direct antitumor activity
   C. Other biological effects

Agents

1.
   A. Alpha
   B. Beta
   C. Gamma

2.
   A. Antiviral
   B. Antiproliferative
   C. Immunomodulatory

3. alpha

4. T cell growth factor

5.
   A. LAK
   B. TIL

6. Directly kills tumor cells by inducing hemorrhagic necrosis

7. hybridoma technology

8. Specific antibodies that are directed against a single antigenic determinant on the surface of a cell

9. Any two:
   A. Lab for research
   B. Pathology for classification
   C. Imaging tumor mass
   D. Targeting treatment modalities
   E. Purging bone marrow of tumor cells

10. Regulate circulating concentrations of granulocytes and mononuclear phagocytes; maturation and differentiation of hematopoietic cells.

11. Any two:
    A. Decrease myelosuppression
    B. Accelerate recovery after bone marrow transplant
    C. Benefit patients with pancytopenia
    D. Treat infections and parasitic disease

12.
    A. II
    B. IV
    C. I
    D. III

PART III. Nursing Management of the Patient Receiving Biological Response Modifiers

The Drug Development Process

1.
    A. II
    B. III
    C. I

2.
    A. III
    B. I
    C. II

Nurse as Patient Educator

1.
    A. Purpose/goal
    B. Route of administration
    C. Benefits and risks associated
    D. Available alternative treatments
    E. Potential side effects

2. Any three:
   A. Special requirements
   B. Financing of clinical trial
   C. Side effects and self-care measures
   D. Self-administration techniques

The Nurse as Patient Advocate

1. patient, health care team

2. support

Continuing Nursing Education Credit

To earn 6.0 hours of continuing nursing education credit, nurse readers should study this book thoroughly, then complete the Continuing Education Test below, recording answers on the Registration Form on the last page of this section. A passing score of at least 70 percent must be attained in order to earn the credit. Nurses achieving a passing score will receive a CNE award certificate from the Nursing Outreach Program of the University of Texas M.D. Anderson Cancer Center, Houston. The center is approved as a provider (No. 90-00588) of continuing education in nursing by the Texas Nurses Association, which in turn is accredited as an approver of continuing education in nursing by the Western Regional Accrediting Committee of the American Nursing Association. Detailed instructions for applying for credit are on the Registration Form.

Continuing Education Test

1. All of the following are functions of the immune system *except*

   A. Defense
   B. Hematopoiesis
   C. Homeostasis
   D. Surveillance

2. __________ is the process by which certain cells of the immune system engulf and destroy invading microorganisms and other antigenic particles.

   A. Phagocytosis
   B. Hematopoiesis
   C. Inflammation
   D. Humoral immunity

3. Natural killer cells are a subpopulation of lymphocytes that recognize cancer cells and destroy the cells by __________.

   A. Phagocytosis
   B. Lysis
   C. Pleuradesis
   D. Epistaxis

4. The hallmark(s) of the adaptive immune system is (are) (a) specificity, (b) memory, (c) cytotoxicity. (Choose one of the following answers.)

   A. (a) only
   B. (b) only
   C. (a) and (b)
   D. all of the above

5. Which of the following is true about IgM?

   A. Functions in hypersensitivity reactions
   B. Found in large amounts during the initial encounter with an antigen
   C. Found in seromucinous secretions
   D. Functions during the secondary immune response

6. Cytotoxic T cells

   A. Decrease the response of T and B cells
   B. Circulate in a "resting state"
   C. Kill cells by lysis
   D. Help initiate the immune response

7. Of the following, which is least likely to affect immune function?

   A. Age
   B. Heredity
   C. Nutritional status
   D. Gender

8. Biological agents are classified according to

   A. Chemical structure
   B. Mode of action
   C. Function during the cell cycle
   D. Side effects

9. Interferons are a family of glycoproteins. Which of the following is *not* a type of interferon?

   A. Alpha interferon
   B. Beta interferon
   C. Epsilon interferon
   D. Gamma interferon

10. One of the ways that interferons exert their anticancer effect is by

    A. Blocking cell surface receptors
    B. Immunomodulatory effects
    C. Cytotoxic mechanisms
    D. Causing free radical formation with cancer cells

11. GM-CSF stimulates growth of

    A. Neutrophils and monocytes
    B. B cells
    C. Monocyte progenitors
    D. All hematopoietic lineages

12. Side effects common to most biological agents include

    A. Fever, chills, and myalgias
    B. Fever, chills, and mucositis
    C. Myalgias, mucositis, and anaphylaxis
    D. Fever, mucositis, and anaphylaxis

13. An optimal plan of care for a patient with fatigue related to biotherapy may include all of the following *except*

    A. Assessment of the effect of fatigue on activities of daily living
    B. Encouraging the patient to arrange the most strenuous activities according to peak energy level
    C. Encourage the patient to request assistance with activities of daily living
    D. Encourage daily aerobic exercise

14. Cytokines are messengers secreted by cells of the

    A. Endocrine system
    B. Immune system
    C. Nervous system
    D. Integument

15. Potential uses of monoclonal antibodies include (a) diagnosis of cancer, (b) symptom management, (c) treatment of cancer. (Choose one of the following answers.)

    A. (a) only
    B. (b) only
    C. (a) and (c)
    D. all of the above

16. Colony-stimulating factors are a family of proteins involved in

    A. Regulation of fever
    B. Controlling the growth of bacterial colonies
    C. The differentiation and maturation of hematopoietic cells
    D. Communication in the nervous system

17. Which of the following clinical trial phases are used to determine response rate for a specific type or stage of disease?

    A. Phase 1
    B. Phase 2
    C. Phase 3
    D. Phase 4

18. Key aspects of teaching patients self-administration of biologics include *all* except

    A. Sterile technique
    B. Scarification
    C. Medication reconstitution
    D. Site selection and rotation

19. In assessing fever in patients receiving biological response modifiers, the most important parameter is

    A. How high the fever is
    B. Whether the fever is in the morning or the evening
    C. The relationship of the fever to the patient's normal response to the BRM
    D. Fever is never of concern in patients receiving BRMs because it is an expected side effect.

20. Allergic reactions are most commonly seen in patients receiving

    A. Interleukins
    B. Interferons
    C. Monoclonal antibodies
    D. Colony-stimulating factors

CNE Registration Form

To keep this book intact, please photocopy this form.

Please rate the book by circling the appropriate answer:

1—poor 2—fair 3—average 4—good 5—excellent

Did the content achieve its objectives?	1	2	3	4	5
Was the content relevant to the objectives?	1	2	3	4	5
Was the content pertinent to your occupation and setting?	1	2	3	4	5
Were you able to meet your personal objectives?	1	2	3	4	5
How long did it take you to study the book?	__________				

Comments:

Record your answers to the 20 questions of the preceding Continuing Education Test by circling the corresponding letters here:

1.	A B C D	8.	A B C D	15.	A B C D
2.	A B C D	9.	A B C D	16.	A B C D
3.	A B C D	10.	A B C D	17.	A B C D
4.	A B C D	11.	A B C D	18.	A B C D
5.	A B C D	12.	A B C D	19.	A B C D
6.	A B C D	13.	A B C D	20.	A B C D
7.	A B C D	14.	A B C D		

This form must be filled out completely for you to receive proper credit of 6.0 contact hours. Please print clearly, or type. (If you wish to retain your book intact, you may photocopy the form.) Enclose check for $12 made payable to M.D. Anderson Nursing Outreach Program and mail to:

M.D. Anderson Cancer Center
Nursing Outreach Program—Box 082
1515 Holcombe Boulevard
Houston, Texas 77030

Check one: ☐ RN ☐ LVN/LPN

State issuing license __________ License number __________

Name ____________________

Street ____________________

City _______________ State _____ ZIP _______

Telephone _____/_________

Index

S

T

U

V